AF854388

The Lost Blueprint of Civilization Forgotten Ideas That Still Shape Modern Life

Mindy Sanchez

The Lost Blueprint of Civilization
Forgotten Ideas That Still Shape Modern Life

The Lost Blueprint of Civilization

Forgotten Ideas That Still Shape Modern Life

By: Mindy Sanchez

Disclaimer

The Lost Blueprint of Civilization explores historical events, archaeological discoveries, scientific research, and cultural traditions from a wide range of reputable sources available at the time of writing. While every effort has been made to present information accurately and responsibly, history is an evolving field. New discoveries, revised interpretations, and ongoing research may expand or challenge current understanding.

This book is intended to encourage curiosity, critical thinking, and a deeper appreciation for humanity's shared past. Some chapters examine long-debated mysteries, competing historical theories, and lesser-known discoveries. When discussing these subjects, the goal is to present evidence, context, and multiple perspectives rather than promote speculation as established fact.

The author does not claim to possess exclusive knowledge or definitive answers regarding unresolved historical questions. Readers are encouraged to explore additional scholarly works, archaeological findings, and primary sources to develop their own informed perspectives.

This book is provided for educational and informational purposes only. It should not be interpreted as academic, legal, scientific, or professional advice. References to individuals, civilizations, cultures, and historical events are made with respect and with the intention of promoting learning and understanding.

Every civilization has contributed to the remarkable story of humanity. If this book inspires you to ask new questions, seek new evidence, and continue exploring the fascinating history of our world, then it has fulfilled its purpose.

Introduction

History is often presented as a collection of dates, famous names, and monumental events. We learn about empires that rose and fell, explorers who crossed oceans, inventors who changed industries, and leaders whose decisions shaped nations. Yet beneath these familiar stories lies another history—one that is quieter, more intricate, and filled with remarkable people, forgotten ideas, and discoveries that rarely make their way into textbooks.

The story of civilization is far richer than many of us realize. Across thousands of years, countless innovations appeared, disappeared, and were rediscovered. Entire cities flourished before fading into the landscape. Brilliant minds introduced ideas that transformed their communities but were later overshadowed by more celebrated figures. Ancient engineers solved problems that still impress modern experts, while cultures separated by vast distances often arrived at surprisingly similar solutions to the same challenges.

This book is an invitation to explore those overlooked chapters of human history.

Rather than focusing only on the events that dominate popular narratives, these pages uncover the connections, innovations, and forgotten achievements that quietly influenced the world we know today. You will encounter remarkable civilizations, little-known scientific breakthroughs, hidden trade networks, misunderstood archaeological discoveries, and individuals whose contributions deserve far greater recognition.

Some stories may surprise you. Others may challenge assumptions you have held for years. Many reveal that history is not a fixed collection of facts but a constantly evolving conversation shaped by new evidence, fresh perspectives, and continuing exploration. Every excavation, every translated manuscript, and every scientific advancement has the potential to reshape what we believe about the past.

Curiosity has always been one of humanity's greatest strengths. It drives exploration, fuels innovation, and encourages us to question what seems certain. Throughout history, many of the most important discoveries began with someone asking a simple question: *What if there is more to the story?*

That same question inspired this book.

As you move through these chapters, you will discover how forgotten knowledge continues to influence modern society in unexpected ways. You will see how ancient ideas echo through today's technologies, how overlooked cultures helped shape global civilization, and how even the smallest discoveries can alter our understanding of history.

The purpose of this book is not to replace established scholarship but to complement it by shining a light on fascinating subjects that deserve greater attention. Every chapter is an opportunity to look beyond the familiar and appreciate the complexity, creativity, and resilience that have defined humanity across the ages.

The past is not as distant as it appears. It lives in our cities, our languages, our inventions, our traditions, and even in the questions we continue to ask. By uncovering these forgotten pieces of history, we gain a richer understanding not only of those who came before us but also of ourselves.

Welcome to a journey through the hidden foundations of civilization. Some of history's greatest stories are the ones that have yet to be fully told.

Chapter 1
The Civilization That Vanished Without a Trace

Imagine waking one morning to discover that an entire nation had disappeared. Its cities stood empty. Its roads stretched into silence. Its monuments weathered under an indifferent sky. There were no survivors to tell the story, no detailed records explaining what had happened, and no clear answer to why a thriving civilization simply ceased to exist. While this sounds like the beginning of a work of fiction, history contains many examples of societies that rose to extraordinary heights before fading into mystery.

Human civilization has never followed a straight path. Throughout history, countless cultures have flourished through innovation, trade, engineering, agriculture, and artistic achievement. Some endured for thousands of years, while others seemed to vanish almost overnight. In many cases, their disappearance was so complete that later generations forgot they had ever existed. Forests reclaimed their cities, deserts buried their roads, and oceans concealed their ports beneath layers of sediment.

One of the greatest misconceptions about history is that civilizations disappear because of a single catastrophic event. While wars, volcanic eruptions, earthquakes, and pandemics have certainly played their part, modern archaeology tells a far more complex story. Most civilizations declined gradually, often through a combination of environmental changes, political instability, economic collapse, shifting trade networks, resource depletion, disease, and social unrest. These pressures accumulated over decades or even centuries until the society could no longer sustain itself.

The ancient city of Mohenjo-daro, once part of the remarkable Indus Valley Civilization, stands as a powerful example. More than four thousand years ago, its carefully planned streets, sophisticated drainage systems, standardized bricks, and advanced urban design rivaled many cities built thousands of years later. Yet despite its impressive achievements, the civilization slowly declined. Researchers now believe that changing river systems, prolonged droughts, and disruptions to agriculture and trade may have played significant roles in its collapse rather than a single dramatic invasion, as once believed.

Across the Atlantic, the Maya civilization reached extraordinary heights in mathematics, astronomy, architecture, and written language. Their towering temples still dominate the jungles of Central America, reminding visitors of a culture capable of remarkable intellectual accomplishments. Yet many of the great southern cities experienced dramatic population declines during what historians call the Classic Maya Collapse. Modern research suggests that severe droughts, political conflict, warfare among competing city-states, environmental degradation, and food shortages combined to create a prolonged period of instability rather than one sudden catastrophe.

Another fascinating mystery lies on Easter Island, known to its Indigenous inhabitants as Rapa Nui. The island's enormous stone statues, called moai, have fascinated explorers for centuries. Early theories blamed environmental destruction caused by the island's inhabitants, but more recent evidence paints a more nuanced picture. Researchers now believe that European diseases, colonial exploitation, forced labor, and outside influences played substantial roles in the population's decline after European contact, demonstrating how historical interpretations can evolve as new evidence emerges.

Not every vanished civilization disappeared because its people died. Sometimes they simply moved. Climate change altered rivers, coastlines shifted, fertile farmland became barren, or trade routes changed direction. Communities migrated in search of better opportunities, leaving behind magnificent cities that slowly surrendered to nature. Generations later, explorers stumbled upon these abandoned places and wondered who had built them.

Nature is remarkably efficient at erasing the footprints of humanity. In tropical regions, dense vegetation can conceal entire cities within a few decades. Tree roots break apart stone walls, vines engulf temples, and layers of soil gradually bury buildings beneath the forest floor. Modern technologies such as LiDAR—Light Detection and Ranging—have transformed archaeology by allowing researchers to scan landscapes from the air. Laser pulses penetrate forest canopies and reveal hidden structures invisible to the naked eye. Entire networks of roads, reservoirs, terraces, and settlements have been discovered beneath jungles that were once believed to be untouched wilderness.

These discoveries remind us that history remains incomplete. Every year, archaeologists uncover forgotten settlements beneath farmland, deserts, glaciers, and even modern cities. Ancient ports lie submerged beneath rising seas, while long-buried manuscripts continue to reveal details about cultures once thought lost forever. Each discovery challenges assumptions and expands our understanding of humanity's past.

Why do civilizations disappear while others endure? The answer appears to lie in adaptability. Societies that successfully respond to environmental changes, economic challenges, political transitions, and external threats often survive periods of crisis. Those unable to adapt become increasingly vulnerable. History demonstrates that resilience is not determined by wealth, military strength, or technological advancement alone. Flexibility, cooperation, and sustainable resource management frequently prove just as important.

There is another lesson hidden within these vanished worlds. Every civilization believed, at least for a time, that its achievements would endure indefinitely. Monumental architecture, thriving economies, and powerful governments created an illusion of permanence. Yet history repeatedly reminds us that no society is immune to change. Even the greatest empires eventually face challenges they never anticipated.

The disappearance of ancient civilizations should not be viewed only as stories of failure. Their achievements continue to influence modern engineering, agriculture, governance, astronomy, mathematics, and urban planning. Although many of their cities fell into ruin, their knowledge survived through cultural exchange, rediscovery, and the enduring curiosity of future generations.

Perhaps the most remarkable truth is that many lost civilizations are not truly lost at all. They have simply been waiting—beneath forests, beneath deserts, beneath oceans, and beneath our assumptions—for someone to uncover their stories once again.

As archaeologists continue to explore the world's hidden landscapes using technologies unimaginable just a generation ago, one thing becomes increasingly clear: history still holds countless secrets. The map of human civilization is far from complete, and some of its most fascinating chapters may still be buried beneath our feet, waiting for the next discovery to rewrite what we think we know.

Chapter 2
Forgotten Cities Beneath Modern Streets

Every bustling city has a story to tell, but some stories are buried beneath layers of pavement, concrete, and steel. Millions of people walk to work, shop, or ride public transportation every day without realizing they are traveling above the remains of earlier civilizations. Hidden beneath modern streets are forgotten neighborhoods, ancient temples, marketplaces, homes, tunnels, and entire cities that once echoed with the sounds of daily life.

Urban landscapes are often thought of as symbols of progress, replacing the old with the new. In reality, cities rarely erase their past completely. Instead, they build upon it. Fires, floods, earthquakes, wars, economic changes, and the simple passage of time lead communities to rebuild rather than relocate. Generation after generation, new streets are laid over old ones, new buildings rise where others once stood, and history becomes sealed beneath layers of development.

This process, known as urban stratification, creates what archaeologists describe as a vertical timeline of human history. Just as geological layers reveal Earth's past, cities preserve centuries—or even thousands of years—of human activity beneath their foundations. Digging only a few feet below street level in some places can uncover evidence of lives lived hundreds or thousands of years ago.

One of the most remarkable examples is Rome. Modern Rome is a vibrant capital filled with traffic, cafés, and historic landmarks, yet beneath its streets lies another Rome entirely. Ancient roads, aqueducts, bathhouses, homes, marketplaces, and temples remain preserved underground. Some structures are so well protected by layers of soil that colorful frescoes, mosaics, and inscriptions have survived for centuries. Archaeologists continue to uncover new sections of the ancient city during routine construction projects, demonstrating that Rome still holds many untold stories.

A similar phenomenon exists in Istanbul, a city that has served as the capital of multiple empires. Beneath modern neighborhoods are the remains of Byzantine churches, Roman forums, Ottoman structures, and massive underground cisterns designed to store water for millions of residents. The famous Basilica Cistern, supported by hundreds of towering marble columns, remained hidden from much of the outside world for centuries before being rediscovered and restored.

Mexico City presents another extraordinary example of history buried beneath progress. The modern metropolis stands atop the ancient Aztec capital of Tenochtitlan, once one of the largest cities in the world. Spanish conquerors dismantled many of its temples and palaces, using their stones to construct colonial buildings. Today, excavations continue to reveal ceremonial platforms, sculptures, canals, and artifacts that offer glimpses into the sophisticated society that flourished there long before European arrival.

London also contains an invisible city beneath its busy streets. Construction of subway lines, office buildings, and transportation systems has uncovered Roman roads, medieval cemeteries, Viking artifacts, forgotten rivers, and entire neighborhoods lost to disasters such as the Great Fire of 1666. Each excavation adds another piece to the city's complex history, reminding residents that London has been continuously rebuilt for nearly two thousand years.

Sometimes entire communities disappear not because they were conquered, but because nature slowly consumed them. Along coastlines, rising sea levels have submerged ports and settlements beneath the ocean. In river valleys, repeated flooding buried villages beneath layers of sediment. Earthquakes have caused cities to sink, while volcanic eruptions have preserved others beneath ash, creating remarkable archaeological time capsules.

One of the most famous examples is Pompeii. When Mount Vesuvius erupted in AD 79, volcanic ash buried the thriving Roman city within hours. Although the eruption claimed countless lives, the ash also preserved homes, artwork, shops, food, and everyday objects with extraordinary detail. Today, Pompeii provides one of the clearest snapshots of daily life in the ancient world, offering insights that written records alone could never provide.

Modern technology has transformed the search for buried cities. Ground-penetrating radar allows researchers to identify walls, foundations, roads, and underground chambers without disturbing the surface. Satellite imagery reveals subtle changes in vegetation and soil that often indicate hidden structures below. LiDAR technology has exposed entire urban landscapes beneath dense forests, while three-dimensional digital mapping enables archaeologists to reconstruct ancient neighborhoods with remarkable precision.

These innovations have fundamentally changed archaeology. Excavation is no longer based solely on educated guesses. Researchers can now detect evidence of long-lost structures before removing a single layer of soil, allowing them to preserve fragile sites while uncovering new chapters of human history.

Unexpected discoveries often occur during ordinary construction projects. Workers digging foundations for apartment buildings, installing water lines, or expanding subway systems occasionally uncover ancient cemeteries, forgotten roads, or buried homes. In many cities, construction must temporarily stop while archaeologists investigate newly discovered remains, balancing modern development with the preservation of cultural heritage.

These discoveries reveal something profound about civilization itself. Cities are not fixed places; they are living organisms that constantly evolve. Every generation reshapes its surroundings while leaving traces for future generations to uncover. What seems permanent today may one day become another hidden layer beneath a future city.

There is also an important lesson in what lies beneath our feet. History is not confined to museums or distant ruins. It exists beneath sidewalks, parking lots, office towers, and public squares. The places we consider modern are often built upon centuries of human ambition, resilience, innovation, and change.

Many of the world's greatest archaeological discoveries have occurred not in remote deserts or isolated jungles but in the heart of thriving cities. Every excavation reminds us that the past has never truly disappeared. It has simply been covered by the next chapter of human progress.

As urban populations continue to grow and cities expand, more forgotten worlds will almost certainly emerge from beneath the streets. Each discovery adds another layer to our understanding of civilization, proving that beneath the foundations of today's world rests an older world, patiently waiting to tell its story.

The next time you walk through a busy city, consider what may lie beneath your feet. A forgotten marketplace. An ancient home. A long-buried temple. Perhaps even the remains of a civilization that once believed its own streets would last forever. History is closer than we often imagine—it is simply hidden below the surface, waiting for someone to uncover it.

Chapter 3
Ancient Engineering Beyond Its Time

Walk through the ruins of an ancient civilization, and one question quickly comes to mind: **How did they build this?** Massive stone blocks weighing hundreds of tons stand perfectly aligned. Aqueducts stretch for miles with carefully calculated gradients. Roads laid thousands of years ago remain visible, while many modern highways require constant repair. Across the globe, ancient engineers created structures so durable and sophisticated that they continue to challenge our understanding of what was possible with the tools available at the time.

Engineering is often viewed as a modern discipline powered by computers, machinery, and advanced materials. Yet long before electricity, steel cranes, or digital blueprints existed, civilizations developed ingenious methods to solve complex problems. Their achievements were not the result of mysterious technologies but of careful observation, mathematical precision, skilled craftsmanship, and generations of accumulated knowledge.

One of the most enduring examples is the Roman Empire. Roman engineers transformed the ancient world by building an extensive network of roads, bridges, aqueducts, and public buildings. Their roads were constructed in multiple layers, beginning with compacted soil, followed by gravel, crushed stone, and fitted paving stones. This layered design improved drainage and stability, allowing many Roman roads to survive for nearly two thousand years.

Perhaps even more remarkable were the Roman aqueducts. Using only gravity, engineers transported millions of gallons of fresh water across valleys, hills, and mountains. Maintaining the correct slope was essential. Too steep, and the water would erode the channels; too shallow, and it would stagnate. Some aqueducts descended only a few feet over several miles, demonstrating extraordinary surveying skills long before laser levels or satellite positioning systems.

Across the Mediterranean, the ancient Egyptians achieved engineering feats that continue to inspire admiration. The Great Pyramid of

Giza, built more than 4,500 years ago, contains an estimated 2.3 million limestone and granite blocks, many weighing several tons. Despite centuries of study, scholars continue to debate the exact methods used to transport and position these enormous stones. Most researchers agree that teams of skilled laborers, carefully designed ramps, sledges, and coordinated planning made the construction possible. Rather than diminishing the achievement, these explanations highlight the remarkable organizational abilities of ancient Egyptian society.

The precision of the pyramid itself is equally astonishing. Its sides are closely aligned with the cardinal directions, and its foundation remains impressively level despite the limitations of the era. Such accuracy required advanced knowledge of geometry, astronomy, and surveying techniques.

Thousands of miles away, the Inca civilization demonstrated a completely different approach to engineering. In the rugged Andes Mountains of South America, Inca builders constructed roads, terraces, bridges, and cities capable of withstanding earthquakes that have destroyed many newer buildings. Their secret lay in flexible construction methods. Instead of relying on mortar, they shaped stones so precisely that they fit together like pieces of an intricate puzzle. During earthquakes, the stones could shift slightly before settling back into place, reducing structural damage.

Machu Picchu serves as one of the finest examples of this engineering philosophy. Built high in the Andes, the city includes sophisticated drainage systems, retaining walls, and agricultural terraces that stabilize the mountainside while preventing erosion. Even after centuries of heavy rainfall, much of its infrastructure continues to function remarkably well.

Ancient engineering also flourished in regions often overlooked in traditional histories. The Indus Valley Civilization, which thrived more than four thousand years ago, developed cities with carefully planned street grids, standardized brick sizes, and one of the world's earliest urban sanitation systems. Houses commonly featured private bathing areas connected to covered drainage networks, reflecting an advanced understanding of public health and urban planning.

In China, engineers constructed sections of the Great Wall across mountains, deserts, and grasslands using locally available materials such as stone, tamped earth, brick, and wood. Rather than serving as a single continuous wall built all at once, the structure evolved over centuries, adapting to changing military strategies and regional landscapes. Its construction required immense logistical planning, resource management, and coordination across vast distances.

Water management was another area in which ancient civilizations excelled. The Nabataeans, whose capital was Petra, transformed a harsh desert environment into a thriving city through an intricate system of channels, reservoirs, dams, and cisterns. They captured seasonal rainfall, reduced flooding, and stored water for dry periods, demonstrating an understanding of hydraulic engineering perfectly suited to their environment.

Similarly, the Khmer civilization developed one of history's most sophisticated water management systems around Angkor. An extensive network of canals, reservoirs, and embankments regulated seasonal floods while supporting agriculture for a vast population. Modern researchers believe this complex infrastructure contributed significantly to the empire's prosperity, though changing climate patterns may have eventually strained the system beyond its capacity.

Ancient engineers also displayed remarkable ingenuity in materials science. Roman builders developed a form of concrete that has survived marine environments for nearly two millennia. Unlike many modern concretes that deteriorate over time, Roman marine concrete appears to strengthen as seawater interacts with its volcanic components. Scientists continue to study this material, hoping its composition may inspire more durable and sustainable construction techniques today.

Another often-overlooked innovation was the widespread use of standardized measurements. Many civilizations established consistent units for bricks, stones, roads, and architectural components, making construction more efficient and ensuring compatibility across large projects. Standardization remains one of the cornerstones of modern engineering, yet its origins stretch back thousands of years.

These achievements remind us that innovation does not always require advanced technology. Ancient engineers succeeded because they carefully observed natural forces, experimented with materials, learned from failures, and passed knowledge from one generation to the next. Their understanding of gravity, water flow, structural stability, and environmental adaptation emerged through experience rather than computer simulations.

Modern engineers continue to study ancient structures not simply out of historical curiosity but because many of their solutions remain relevant. Sustainable construction, earthquake-resistant design, passive cooling systems, efficient water management, and durable building materials are all areas where ancient practices offer valuable insights. As climate change and rapid urbanization challenge today's infrastructure, lessons from the past are becoming increasingly important.

Perhaps the greatest misconception about ancient engineering is that it belonged to a primitive world. On the contrary, these civilizations demonstrated extraordinary intelligence, creativity, and practical problem-solving using the resources available to them. Their accomplishments remind us that technological progress is not measured solely by machines but by the ability to understand nature and work with it.

The monuments that have survived for centuries are more than historical landmarks. They are enduring evidence of human ingenuity, perseverance, and imagination. Every carefully fitted stone, every precisely measured channel, and every towering structure tells a story of people who refused to accept the limits of their time.

Long before blueprints were printed or cranes filled the skyline, ancient engineers proved that remarkable achievements begin not with technology, but with vision, knowledge, and the determination to build something that would outlast generations.

Chapter 4
The Libraries That Changed History

Before the internet connected the world, before printing presses made books widely available, and before digital archives stored humanity's knowledge in seconds, libraries served as the guardians of civilization's memory. They were more than buildings filled with written texts. They were centers of discovery, education, debate, scientific advancement, and cultural preservation.

Throughout history, some libraries became symbols of human curiosity and ambition. Within their walls, scholars studied the movements of planets, translated ancient languages, recorded medical knowledge, preserved literature, and exchanged ideas that shaped the future. Some libraries survived for centuries, while others disappeared through war, political upheaval, natural disasters, or simple neglect. Yet their influence continued long after their physical collections were lost.

The story of civilization is, in many ways, a story of humanity's effort to preserve knowledge. Every manuscript copied, every scroll protected, and every discovery recorded represented a belief that ideas should outlive the people who created them.

One of the most famous examples is the Library of Alexandria, a legendary center of learning in ancient Egypt. Established during the Hellenistic period, it became associated with one of the greatest intellectual ambitions of the ancient world: gathering knowledge from across different cultures and making it available to scholars.

The library was part of a larger research institution known as the Mouseion, which attracted philosophers, mathematicians, scientists, poets, and historians. Scholars connected to Alexandria contributed to fields such as astronomy, geometry, geography, medicine, and literature. They studied texts from Egypt, Greece, Persia, and other regions, creating an environment where ideas from different civilizations could meet and evolve.

The exact fate of the Library of Alexandria remains debated. Popular stories describe a single dramatic destruction event, but historians generally believe its decline occurred gradually through a combination of political changes, conflicts, reduced funding, and the shifting importance of Alexandria over time. Whether destroyed in one moment or lost through centuries of decline, the disappearance of its collection represents one of history's greatest losses.

What makes the story of Alexandria so powerful is not only what was lost, but what it represented. It reflected the belief that knowledge from different cultures could be collected, studied, and shared. That idea continues to influence libraries and research institutions around the world today.

Far from Alexandria, another remarkable center of knowledge emerged in ancient Mesopotamia. The Library of Ashurbanipal, located in the Assyrian capital of Nineveh, contained thousands of clay tablets written in cuneiform script. Unlike fragile scrolls, these tablets were preserved through the unusual durability of baked clay.

The collection included myths, historical records, legal documents, medical texts, astronomical observations, and religious writings. Among its most famous discoveries was a version of the **Epic of Gilgamesh**, one of humanity's earliest known literary works. The library provided modern researchers with invaluable information about ancient languages, beliefs, and daily life.

The preservation of these tablets demonstrates how different civilizations developed unique methods for protecting knowledge. While some cultures relied on papyrus, parchment, or paper, the people of Mesopotamia created records that survived thousands of years because of the very material on which they were written.

Another extraordinary example comes from the ancient Buddhist world: Nalanda University in India. Established around the fifth century CE, Nalanda became one of the world's earliest residential centers of higher learning. Students and scholars traveled from distant regions to study philosophy, medicine, mathematics, astronomy, and religious texts.

Nalanda's libraries were reportedly among the largest collections of written knowledge in the ancient world. Manuscripts covered multiple subjects and reflected centuries of intellectual exchange across Asia. The institution represented something rare for its time: a place where education, research, and international collaboration existed together.

The destruction of Nalanda in the twelfth century marked a devastating loss for scholarship in the region. Thousands of manuscripts were destroyed, and a major center of learning disappeared. However, its influence continued through the scholars it educated and the knowledge they carried elsewhere.

Libraries also played a significant role during the Islamic Golden Age. In Baghdad, the House of Wisdom became a major center of translation, research, and intellectual exchange during the Abbasid period. Scholars translated works from Greek, Persian, Indian, and other traditions into Arabic, preserving and expanding knowledge in mathematics, astronomy, medicine, philosophy, and geography.

Many scientific advancements associated with this period were built upon earlier discoveries while introducing new ideas and methods. Scholars improved mathematical systems, developed astronomical instruments, wrote medical encyclopedias, and expanded geographic understanding. The preservation and advancement of knowledge during this era later influenced European scholarship.

Libraries were never simply storage spaces. They were engines of innovation. A single manuscript could inspire generations of thinkers. A translated text could introduce an entire civilization to new concepts. A preserved observation could become the foundation for future scientific breakthroughs.

The loss of libraries throughout history also reveals a painful truth: knowledge is fragile. Wars, political conflicts, censorship, and disasters have repeatedly threatened humanity's written heritage. When a library disappears, the loss extends beyond books. It can mean the disappearance of languages, personal stories, scientific discoveries, artistic traditions, and records of ordinary lives.

Yet history also shows humanity's determination to recover what has been lost. Archaeologists continue discovering forgotten texts buried beneath ruins. Scholars reconstruct damaged manuscripts. Ancient languages once considered impossible to understand have been deciphered. Each recovery restores a small piece of the human story.

Modern libraries continue this ancient mission in new forms. Digital archives, universities, museums, and research centers now preserve information on a scale unimaginable to earlier civilizations. Yet the fundamental purpose remains unchanged: protecting knowledge so future generations can learn, question, and create.

The greatest libraries in history changed the world not because of their buildings, but because of the ideas they protected. They connected people across centuries and continents, proving that knowledge does not belong to a single generation. It is a shared inheritance passed from one era to the next.

Every book, manuscript, inscription, and digital record represents humanity's ongoing conversation with itself. The libraries that changed history remind us that civilizations are remembered not only by the monuments they built, but by the knowledge they chose to preserve.

Because when knowledge survives, humanity's discoveries, dreams, and questions continue long after the original voices have disappeared.

Chapter 5
Lost Languages Waiting to Be Deciphered

Every civilization leaves behind a voice. Sometimes that voice survives through monuments, tools, artwork, or architecture. But the most direct connection to the people of the past comes through their words. Languages reveal how civilizations understood the world, organized their societies, expressed their beliefs, recorded their histories, and shared their dreams.

Yet countless languages have disappeared throughout human history. Some vanished because entire communities declined. Others were replaced by more dominant languages through conquest, migration, or cultural change. In many cases, the written symbols remain, carved into stone or preserved on ancient tablets, but their meanings have been lost.

These forgotten languages represent some of archaeology's greatest mysteries. They are messages waiting across thousands of years, carrying information from people who can no longer explain their own words. Each successful decipherment is like opening a sealed doorway into another world.

The study of lost languages has transformed our understanding of history. Civilizations once considered mysterious have become more familiar because researchers learned how to read their writings. Ancient laws, personal letters, religious beliefs, trade records, and scientific observations have all emerged from languages that were once silent.

One of the greatest breakthroughs in language decipherment came with the ancient Egyptian writing system known as hieroglyphs. For centuries, the magnificent inscriptions covering Egyptian temples and monuments remained unreadable to scholars. Many researchers recognized that these symbols represented a sophisticated writing system, but the key to understanding them had been lost.

The discovery of the Rosetta Stone in 1799 changed everything. The stone contained the same decree written in three scripts: Greek, Egyptian demotic, and Egyptian hieroglyphs. Because scholars could already read Greek, they used it as a guide to unlock the Egyptian text. The work of deciphering the inscriptions, especially by scholars such as Jean-François Champollion, opened a new era of Egyptology.

Suddenly, ancient Egypt could speak again. Pharaohs were no longer only names carved into monuments; they became historical figures with ambitions, conflicts, religious beliefs, and personal stories. Tomb inscriptions revealed ideas about death and the afterlife. Medical texts provided insight into ancient treatments. Administrative records showed how a complex society managed agriculture, taxation, and government.

Another remarkable achievement involved the decipherment of cuneiform, one of the world's earliest writing systems. Developed in ancient Mesopotamia, cuneiform was created by pressing wedge-shaped marks into clay tablets. Over thousands of years, this writing system was adapted by different civilizations, including the Sumerians, Akkadians, Babylonians, and Assyrians.

For a long time, these mysterious symbols seemed impossible to understand. The breakthrough came through the combined efforts of many scholars who compared inscriptions, studied repeated patterns, and used multilingual texts as clues. Once deciphered, cuneiform revealed an extraordinary archive of human experience.

Clay tablets preserved everything from royal announcements to ordinary business transactions. They recorded marriage agreements, property sales, mathematical calculations, astronomical observations, myths, and personal letters. These documents showed that ancient societies were not distant and unfamiliar—they contained people with everyday concerns remarkably similar to our own.

One of the most fascinating undeciphered writing systems is Linear A, created by the Minoan civilization on the island of Crete. The Minoans built impressive palaces, developed extensive trade networks, and produced remarkable artwork. Their writing system appears on clay tablets and inscriptions, but despite decades of study, its language remains unknown.

Researchers have identified patterns within Linear A, but they have not yet discovered enough information to fully translate it. The mystery remains because scholars do not have a bilingual inscription similar to the Rosetta Stone. Without that crucial connection, the symbols continue to guard their secrets.

The Indus Valley Civilization presents another extraordinary challenge. More than four thousand years ago, this civilization created cities with advanced urban planning, sanitation systems, and long-distance trade connections. Archaeologists have discovered thousands of short inscriptions carved onto seals and objects, yet the Indus script has not been successfully deciphered.

The difficulty lies in the limited length of the inscriptions. Most examples contain only a small number of symbols, making it difficult to determine grammatical patterns or vocabulary. Researchers continue to debate whether the script represents a complete writing system, a symbolic system, or something else entirely. Unlocking the Indus script could reveal the names of cities, rulers, beliefs, and daily activities of one of the ancient world's most impressive civilizations.

The mystery of the Etruscan language provides another example of how difficult it can be to recover forgotten voices. The Etruscans influenced early Roman culture through architecture, religion, art, and political traditions. Their alphabet can be read because it was based on Greek writing, but much of the language itself remains poorly understood because relatively few texts survived.

Some lost languages disappear completely, leaving behind only traces. Others survive through unexpected discoveries. A forgotten inscription on a stone wall, a buried tablet, or a damaged manuscript can provide the missing clue needed to unlock centuries of silence.

Modern technology is changing the search for lost languages. Digital imaging allows researchers to examine faded inscriptions without damaging them. Artificial intelligence helps identify patterns across thousands of texts. Computational linguistics allows scholars to compare ancient writing systems with known languages and test possible interpretations.

However, technology alone cannot solve every mystery. Decipherment requires patience, creativity, and a deep understanding of history. Researchers must consider cultural context, geography, trade connections, and relationships between civilizations. A symbol may represent a sound, an idea, a word, or something entirely different depending on the writing system.

The search for lost languages is more than an academic pursuit. Language is the foundation of human memory. When a language disappears, an entire way of understanding the world can disappear with it. Words carry traditions, humor, beliefs, emotions, and knowledge that cannot always be translated into another language.

Every undeciphered script represents a civilization still waiting to introduce itself. Behind each symbol is a person who wanted to communicate something important enough to record. They wrote about their leaders, their families, their gods, their discoveries, and their daily lives with the hope that their words would endure.

Thousands of years later, those messages remain.

The future may bring discoveries that finally unlock these forgotten voices. A newly discovered inscription, a hidden archive, or a breakthrough in technology could reveal what ancient people were trying to tell us all along.

History is filled with conversations interrupted by time. Lost languages remind us that humanity's story is much larger than the languages we speak today. Across stone, clay, and forgotten symbols, countless voices are still waiting to be heard.

Chapter 6
Hidden Trade Routes That Connected the World

Long before modern highways, shipping containers, airplanes, and digital marketplaces connected the planet, humanity was already building a global network of exchange. Across deserts, mountains, oceans, and forests, ancient travelers carried goods, ideas, technologies, beliefs, and stories from one civilization to another.

Trade routes were not merely paths for moving valuable objects. They were channels through which cultures met and transformed one another. A piece of silk, a precious metal, a spice, or a rare stone could travel thousands of miles, passing through countless hands before reaching its final destination. Along the way, languages mixed, inventions spread, religions expanded, and societies changed.

Many of these ancient routes remained hidden for centuries. Some were buried beneath shifting sands. Others disappeared when political powers collapsed or when new trade networks replaced them. Yet modern archaeology continues to reveal that the ancient world was far more connected than many once believed.

The popular image of the ancient past often portrays civilizations as isolated communities separated by vast distances. The reality was very different. Merchants, explorers, sailors, and migrants created complex networks that linked continents together long before modern globalization.

One of the most famous examples is the Silk Road, a vast network of trade routes that connected East Asia, Central Asia, the Middle East, and Europe. Contrary to its name, the Silk Road was not a single road but a collection of interconnected pathways stretching across thousands of miles.

Beginning in ancient China, silk became one of the most valuable commodities traded along these routes. The luxurious fabric traveled westward through deserts and mountain passes, reaching distant markets where it became a symbol of wealth and prestige. However, silk was only one part of a much larger exchange. Traders carried spices, glassware, precious stones, metals, textiles, medicines, and artistic ideas across continents.

The Silk Road also transported something far more powerful than physical goods: knowledge. Technologies such as papermaking, printing methods, mathematical concepts, and agricultural techniques spread between cultures through these connections. Philosophies and religions traveled alongside merchants, shaping societies thousands of miles away from their origins.

The dangers of these journeys were immense. Traders crossed harsh deserts, freezing mountain ranges, and politically unstable regions. They faced extreme weather, bandits, disease, and unfamiliar landscapes. Despite these challenges, the rewards of long-distance trade encouraged generations of travelers to continue these dangerous journeys.

While the Silk Road connected the eastern and western portions of the ancient world, maritime trade routes created another form of global connection. Ancient sailors learned to navigate vast oceans using knowledge of winds, currents, stars, and seasonal weather patterns.

The Indian Ocean became one of history's greatest trade networks. For thousands of years, merchants traveled between East Africa, Arabia, India, Southeast Asia, and China. Seasonal monsoon winds allowed ships to travel predictable routes, creating a maritime highway that connected distant civilizations.

Through these ocean networks, merchants exchanged spices, ivory, gold, ceramics, textiles, and precious woods. Coastal cities became wealthy centers of commerce, attracting people from different cultures and creating diverse communities where multiple languages and traditions existed side by side.

The ancient city of Aksum, located in present-day Ethiopia and Eritrea, became a powerful trading empire partly because of its strategic position along Red Sea routes. Aksum connected African markets with Arabia, India, and the Mediterranean world. Its merchants traded goods such as ivory, gold, and agricultural products while participating in a broader international economy.

Another remarkable trade network existed across the Sahara Desert. Today, the Sahara is often imagined as an impossible barrier, but for centuries it served as a bridge between North Africa and sub-Saharan Africa. Camel caravans crossed the desert carrying salt, gold, textiles, and other valuable resources.

The introduction of camel transportation transformed desert commerce. Camels could survive long journeys with limited water, allowing merchants to establish routes across one of the harshest environments on Earth. These trade connections contributed to the rise of powerful West African kingdoms, including Ghana, Mali, and Songhai.

Gold from West Africa became especially important in international markets. It traveled north across the Sahara and eventually reached Mediterranean and Middle Eastern economies. In exchange, African kingdoms received salt, manufactured goods, books, and other resources. These exchanges demonstrate that Africa was not isolated from world history but was deeply connected to global networks of commerce and culture.

Trade routes also existed in places where historians once assumed little interaction occurred. Archaeological discoveries in the Americas reveal extensive networks connecting distant communities long before European contact. Indigenous peoples exchanged materials such as obsidian, shells, cacao, feathers, copper, and agricultural products across thousands of miles.

The ancient Maya, for example, participated in complex trade networks throughout Mesoamerica. Coastal communities exchanged marine resources with inland cities, while valuable materials moved between regions with different environments. These connections helped support urban growth and cultural exchange.

One of the most surprising discoveries in recent decades has been evidence of ancient trade reaching even farther than previously imagined. Objects found far from their original sources have forced researchers to reconsider how mobile ancient societies truly were. A rare material discovered hundreds or thousands of miles from its origin raises important questions: Who carried it? How did they travel? What relationships allowed these exchanges to occur?

Trade routes were also pathways for ideas. A merchant carrying goods could unknowingly carry knowledge that would transform another society. Agricultural techniques, artistic styles, medical practices, religious beliefs, and scientific concepts often spread through commercial connections.

However, trade also had darker consequences. The same routes that carried valuable goods and knowledge also allowed diseases to spread between populations with little immunity. Epidemics could travel along the same pathways as merchants and travelers, dramatically affecting societies across continents.

The rise and fall of trade routes often reflected the changing fortunes of civilizations. When empires provided security and maintained infrastructure, commerce flourished. When political instability, warfare, or environmental changes disrupted connections, once-thriving trade centers could decline.

Many ancient trade routes disappeared not because they failed, but because the world changed. New technologies, political boundaries, and economic systems redirected the movement of goods. Ports that once welcomed thousands of ships became quiet ruins. Desert crossroads that once supported thriving communities became abandoned settlements.

Yet these forgotten routes continue to reveal humanity's interconnected past. Every artifact discovered far from its origin tells a story of movement, cooperation, and exchange. A single object can provide evidence of journeys made by people whose names have been lost to history.

The hidden trade routes of the ancient world remind us that globalization did not begin in the modern era. Humanity has always been connected. Long before satellites and shipping networks, people crossed enormous distances searching for resources, opportunities, and relationships.

The world has never been as divided as it appears. Beneath deserts, beneath oceans, and beneath modern cities are the paths that reveal a shared human history—one built not only by rulers and armies, but by ordinary travelers who carried goods, ideas, and stories across the world.

The roads may have disappeared, but their impact remains. The connections created by ancient traders helped shape the languages we speak, the foods we eat, the technologies we use, and the cultures we know today.

The world was connected long before we had a name for it.

Chapter 7
The Original Inventors History Forgot

Every invention has a story. Some begin with a famous name attached to a patent, a laboratory, or a groundbreaking demonstration. Others begin quietly, created by individuals whose names disappeared while their ideas transformed the world.

History often remembers the final person who perfected an invention rather than the many people who contributed to its development. Innovation is rarely the work of one isolated genius. It is usually a long chain of discoveries, experiments, improvements, and adaptations passed from one generation to another.

Behind many of the technologies and systems we rely on today are forgotten inventors whose contributions shaped civilization. Their names may not appear in popular history books, but their ideas continue to influence how humanity lives, communicates, travels, heals, and builds.

The history of invention is not only a story of famous breakthroughs. It is also the story of overlooked minds whose creativity changed the course of human progress.

One of the earliest and most important inventions in human history was agriculture. For thousands of years, hunter-gatherer communities developed methods for cultivating plants, managing animals, and understanding seasonal cycles. These discoveries were not credited to individual inventors because they emerged through collective experimentation over generations.

The people who first selected stronger seeds, learned irrigation techniques, and developed farming systems created the foundation for cities, governments, and civilizations. Without these anonymous innovators, the development of complex societies would have been impossible.

Another forgotten group of inventors were the early engineers of ancient Mesopotamia. The Sumerians created some of humanity's earliest technological systems, including the wheel, writing systems, advanced irrigation methods, and organized urban planning. While later civilizations expanded upon these achievements, the original innovators laid the groundwork for many aspects of modern life.

The wheel is often considered one of humanity's greatest inventions, yet its exact creator remains unknown. It was not simply a matter of placing a circle beneath an object. The development of the wheel required an understanding of rotational movement, materials, balance, and manufacturing. Its invention transformed transportation, agriculture, pottery production, and industry forever.

Ancient inventors also changed how humans measured and understood the world. The Babylonians developed sophisticated mathematical systems that influenced later scientific traditions. Their use of a base-60 number system helped create the foundation for measurements still used today, including the division of hours into minutes and seconds.

In ancient Egypt, unknown architects, craftsmen, and engineers created monumental structures that demonstrated advanced knowledge of mathematics, construction, and organization. While pharaohs receive much of the historical attention, the workers and designers behind these achievements developed techniques that influenced architecture for centuries.

The history of medicine also contains many overlooked pioneers. Ancient healers developed treatments, surgical methods, and observations that formed early medical knowledge. Egyptian physicians documented injuries, diseases, and remedies thousands of years before modern medicine existed. Ancient Indian surgeons developed techniques for reconstructive procedures, while Greek physicians advanced the study of anatomy and diagnosis.

Many of these medical innovators remain anonymous because their knowledge was preserved through traditions rather than personal records. Their discoveries remind us that scientific progress often begins with careful observation and a willingness to test new ideas.

Women have also played a significant role in invention, though their contributions have frequently been overlooked. Throughout history, women developed tools, methods, and discoveries while facing limited access to education, recognition, and ownership of their work.

For example, many women contributed to computing, mathematics, engineering, and scientific research long before these fields became widely accessible. Their work helped shape modern technology, yet their stories were often minimized or attributed to others.

One example is the ancient mathematician and philosopher Hypatia of Alexandria, who contributed to astronomy, mathematics, and scientific teaching during a time when women rarely held public intellectual roles. Although many details of her work were lost, her legacy represents the importance of preserving the contributions of thinkers who challenged the boundaries of their era.

Innovation has also come from unexpected places. Indigenous communities around the world developed technologies perfectly adapted to their environments. They created agricultural systems, navigation methods, medicines, architecture, and ecological practices based on generations of observation.

The Maya developed advanced astronomical calculations and calendar systems. Polynesian navigators crossed vast oceans using knowledge of stars, waves, winds, and marine patterns. Indigenous agricultural techniques, including methods of soil management and crop diversity, continue to influence modern sustainability research.

These achievements demonstrate that invention is not limited to laboratories or industrial centers. Human creativity appears wherever people face challenges and search for solutions.

Many inventors were forgotten because history was often recorded by those with political power. Written records frequently focused on rulers, military victories, and wealthy individuals rather than craftsmen, farmers, builders, healers, and ordinary problem-solvers.

In some cases, inventors were overlooked because their ideas were ahead of their time. Technologies sometimes appear, disappear, and reappear centuries later when society is ready to use them. Ancient designs for mechanical devices, automated systems, and scientific instruments reveal that imagination has always existed—even when technology could not fully support those ideas.

The Antikythera mechanism provides a fascinating example. Discovered in a shipwreck near Greece, this ancient device contained complex gears capable of tracking astronomical cycles. Created more than two thousand years ago, it demonstrated mechanical knowledge far beyond what many historians once expected from the ancient world.

The inventor or inventors behind this remarkable device remain unknown. Their names vanished, but their intelligence survived through the machine they created.

The same pattern appears throughout history: the creator disappears, but the creation remains.

Modern society often celebrates innovation as a sudden moment of inspiration, yet the truth is more complex. Every invention stands upon previous discoveries. The smartphone, for example, depends on centuries of advancements in mathematics, communication, electricity, computing, materials science, and engineering. The final product represents a long history of human creativity.

Recognizing forgotten inventors does more than correct historical records. It changes how we understand progress itself. Innovation is not created by a small number of extraordinary individuals alone. It is built through the combined efforts of countless people whose ideas contribute to a larger human story.

The forgotten inventors of history remind us that every generation inherits knowledge from those who came before. The tools we use, the systems we depend on, and the discoveries we celebrate are connected to countless unnamed minds.

Some inventions are remembered. Some inventors are celebrated. Others disappear into history.

But their ideas remain.

And sometimes, the greatest innovators are the ones whose names we never learned.

Chapter 8
Ancient Medicine That Science Is Rediscovering

For thousands of years, humans have searched for ways to understand the body, treat illness, reduce pain, and extend life. Long before modern hospitals, advanced imaging, and laboratory medicine existed, ancient healers observed nature, studied human anatomy, experimented with treatments, and developed medical practices that reflected a deep understanding of the world around them.

Some ancient medical ideas were inaccurate by modern standards, while others were surprisingly advanced. As science continues to investigate historical practices, researchers are discovering that many traditional remedies and techniques contained valuable knowledge that had been overlooked, misunderstood, or forgotten.

The history of medicine is not simply a story of replacing old ideas with new ones. It is a continuous process of observation, experimentation, and improvement. Modern science often builds upon discoveries made centuries earlier, and some ancient practices are now being examined with advanced technology to understand why they worked.

One of the earliest recorded medical traditions comes from ancient Egypt. Egyptian physicians developed organized approaches to treating injuries, illnesses, and chronic conditions more than four thousand years ago. Medical knowledge was recorded on papyrus scrolls, including descriptions of wounds, surgical procedures, medicines, and observations of the human body.

The Edwin Smith Papyrus, one of the oldest known medical documents, contains detailed discussions of injuries and their treatment. Unlike many ancient medical texts focused on supernatural explanations, this document demonstrates careful observation and a practical approach to trauma care. It describes examining wounds, identifying symptoms, and predicting outcomes based on physical evidence.

Ancient Egyptian healers also developed methods for treating fractures, applying bandages, and using plant-based remedies. Some treatments included ingredients that modern researchers continue to study for their antimicrobial or healing properties.

Ancient Mesopotamian civilizations also contributed significantly to early medicine. Physicians recorded symptoms, diagnoses, and treatments on clay tablets. Their writings reveal attempts to understand disease patterns, identify causes, and create systematic approaches to healing.

Although some Mesopotamian medical practices were connected to spiritual beliefs, they also included practical treatments using herbs, minerals, and physical procedures. These records provide valuable insight into how early societies attempted to understand the relationship between the body, environment, and illness.

In ancient India, the medical tradition known as Ayurveda developed a comprehensive approach to health that emphasized balance, prevention, diet, lifestyle, and natural remedies. Texts associated with Ayurvedic medicine described surgical procedures, anatomy, medicinal plants, and methods for maintaining overall wellness.

One of the most remarkable figures connected to ancient Indian medicine was Sushruta, often associated with early surgical advancements. Historical texts attributed to the Sushruta tradition describe techniques involving wound care, reconstructive procedures, and surgical instruments. Modern researchers have examined these writings to better understand the development of early surgical knowledge.

Ancient Greek medicine introduced another major shift in how people understood disease. Instead of explaining illness only through supernatural causes, Greek physicians increasingly focused on observation and natural explanations.

Hippocrates, often called the "father of medicine," became associated with the idea that physicians should carefully observe symptoms and consider environmental factors when treating patients. The Hippocratic tradition emphasized medical ethics, careful diagnosis, and the importance of documenting observations.

Greek physicians also advanced anatomical knowledge and influenced later medical education. Their writings became foundational texts studied for centuries.

One area where ancient medicine continues to surprise researchers is the use of medicinal plants. Many traditional remedies were based on generations of experimentation with local plants, roots, flowers, and minerals. While not every historical remedy was effective, some contained compounds that modern medicine has adapted or studied.

For example, willow bark was used in traditional healing practices for pain and fever. Centuries later, scientists identified salicin, a compound related to aspirin. The journey from traditional remedy to modern medication demonstrates how historical knowledge can inspire scientific discovery.

Similarly, ancient cultures used certain plants with antimicrobial properties long before scientists understood bacteria. Modern research continues examining traditional medicines to identify compounds that may contribute to future treatments.

Ancient surgery also reveals unexpected sophistication. Archaeological evidence shows that early surgeons performed procedures such as setting broken bones, treating wounds, and conducting operations using specialized tools. Some ancient surgical instruments discovered by archaeologists resemble early versions of tools still used in modified forms today.

The study of ancient remains has also revealed evidence of medical care. Healed fractures show that someone provided treatment and support. Signs of surgery demonstrate that ancient communities possessed knowledge and skills often underestimated by modern observers.

Dental care is another field where ancient practices offer surprising discoveries. Archaeological findings reveal attempts to treat tooth problems, create dental replacements, and maintain oral health. Some ancient cultures used mixtures of natural materials to clean teeth, while others developed early forms of dental repair.

Ancient Chinese medicine also contributed influential ideas about health and the human body. Traditional Chinese practices emphasized balance, prevention, and the relationship between physical and environmental factors. Acupuncture, herbal medicine, and diagnostic techniques have been studied extensively in modern times, with researchers investigating their effectiveness and biological mechanisms.

The rediscovery of ancient medicine does not mean every historical practice should replace modern healthcare. Scientific testing remains essential. Some traditional treatments have proven useful, while others have been shown to be ineffective or potentially harmful. The value lies in investigation—examining old knowledge with modern tools to determine what lessons remain relevant.

Today, researchers use advanced methods to study ancient medical practices. Chemical analysis can identify compounds in archaeological materials. DNA research can reveal ancient disease patterns. Imaging technology allows scientists to examine preserved remains without causing damage.

These discoveries are changing the way we view ancient healers. They were not simply relying on superstition or guesswork. Many were careful observers who documented patterns, tested treatments, and passed knowledge forward.

Perhaps the most important lesson from ancient medicine is the power of curiosity. Early healers faced the same fundamental questions that modern scientists continue to explore: What causes illness? How does the body heal? How can suffering be reduced?

The tools have changed, but the questions remain.

The story of medicine is not a straight line from ignorance to knowledge. It is a long journey filled with discoveries, corrections, forgotten ideas, and renewed exploration. Ancient healers created the foundation upon which later generations built, and their observations continue to inspire scientific research today.

Some of their secrets were lost. Some were misunderstood. Some are now being rediscovered.

The past still has remedies to reveal—not because ancient knowledge was perfect, but because human curiosity has always been searching for answers.

Chapter 9
The Secret Mathematics of the Ancients

Numbers are often considered a modern language of science, technology, and engineering. They power computers, guide spacecraft, shape economies, and help us understand the universe. Yet thousands of years before calculators, digital systems, and advanced mathematics existed, ancient civilizations were already exploring the hidden patterns of numbers.

They measured the movement of stars, calculated the dimensions of buildings, tracked time, predicted seasons, managed trade, and developed mathematical systems that continue to influence the modern world. Much of this knowledge remained hidden for centuries because it was recorded in unfamiliar symbols, ancient languages, and methods that required careful interpretation.

The mathematics of ancient civilizations reveals a surprising truth: humans have always searched for order within the world around them. Numbers were not merely tools for counting. They were a way to understand nature, organize society, and explore the mysteries of existence.

One of the earliest known mathematical traditions emerged in Mesopotamia, where the Sumerians and Babylonians developed advanced systems of calculation thousands of years ago. Their knowledge was recorded on clay tablets using cuneiform writing, preserving equations, measurements, and problem-solving methods.

The Babylonians developed a sophisticated base-60 number system, which may seem unusual today but left a lasting impact. Our modern division of time into sixty minutes per hour and sixty seconds per minute comes from this ancient mathematical tradition.

Their astronomers also used mathematics to track celestial movements with remarkable accuracy. They recorded the cycles of the Moon, calculated planetary positions, and created detailed astronomical tables. These observations helped establish early foundations for later astronomy.

Babylonian mathematics was not limited to astronomy. Clay tablets reveal that they solved complex problems involving geometry, algebra, and practical calculations. They calculated areas, volumes, interest rates, and measurements needed for construction and agriculture.

Far from Mesopotamia, ancient Egyptians developed their own remarkable mathematical approaches. Their civilization depended on measurement. The annual flooding of the Nile River required careful surveying to restore property boundaries and maintain agricultural systems.

Egyptian builders used geometry to create monuments with extraordinary precision. The alignment and proportions of structures such as the pyramids demonstrate advanced knowledge of angles, measurements, and spatial relationships.

The Egyptians also developed practical mathematical methods for architecture, taxation, trade, and engineering. Their mathematical knowledge was not created for abstract theory alone—it was deeply connected to solving real-world problems.

Ancient Greek mathematicians transformed mathematics into a more theoretical discipline. They explored not only how numbers worked but why mathematical principles existed.

The Greek mathematician Pythagoras and his followers studied numerical relationships, geometry, and patterns found throughout nature. The famous theorem associated with Pythagoras became one of the most recognizable principles in geometry, though similar mathematical relationships had been understood in other civilizations before his time.

Euclid, another influential Greek thinker, organized geometric knowledge into a structured system. His work became one of the most influential mathematical texts in history, shaping education for centuries.

The Greeks also explored concepts that pushed the boundaries of ancient thought. They studied infinity, irrational numbers, proportions, and mathematical proofs. Their approach emphasized logical reasoning, creating a foundation for many areas of modern mathematics.

Ancient India contributed some of the most important mathematical ideas in human history. Indian mathematicians developed concepts that transformed global calculation, including the use of zero as a number and the decimal place-value system.

The invention of zero was revolutionary. It provided a way to represent nothing while also allowing numbers to expand through positional notation. This simple idea became essential to modern mathematics, science, computing, and engineering.

Indian mathematicians also made significant advances in algebra, astronomy, and trigonometry. Their discoveries traveled through trade and scholarship networks, eventually influencing mathematical traditions around the world.

The Maya civilization developed another extraordinary mathematical system. In Mesoamerica, Maya scholars created a sophisticated understanding of time, astronomy, and calculation. They developed a place-value system and independently used a concept similar to zero.

Their calendar systems reflected detailed astronomical observations. Maya astronomers tracked the movements of the Sun, Moon, and planets with remarkable precision. Their calculations allowed them to predict celestial events and organize religious and agricultural activities.

What makes ancient mathematics especially fascinating is how many civilizations developed similar ideas independently. Across different continents, separated by oceans and thousands of miles, people discovered ways to measure space, track time, and understand patterns.

This suggests that mathematics is not merely a cultural invention—it is also a response to the universal challenges humans face. Every society must measure resources, organize communities, understand seasons, and solve practical problems.

Ancient mathematics was also closely connected to architecture. The design of temples, pyramids, cities, and monuments required careful planning and measurement. Mathematical principles influenced proportions, symmetry, alignment, and structural stability.

Some ancient structures appear to reflect deep knowledge of geometry and astronomy. While researchers continue to debate the exact intentions behind certain designs, there is no doubt that many ancient builders possessed impressive technical understanding.

Astronomy was one of the greatest motivations for mathematical development. Ancient people looked to the sky not only with wonder but with practical purpose. Predicting seasons helped farmers determine planting cycles. Tracking stars helped travelers navigate. Understanding celestial patterns helped societies organize calendars and religious ceremonies.

The relationship between mathematics and astronomy created some of humanity's earliest scientific traditions.

Ancient mathematical knowledge was not always preserved. Wars, political changes, and the destruction of libraries caused many discoveries to disappear. Some ideas survived only because they were copied, translated, and passed through different cultures.

The rediscovery of ancient mathematical texts has revealed how much knowledge existed long before modern science. Each recovered manuscript challenges the assumption that earlier societies lacked intellectual sophistication.

Today, researchers continue studying ancient mathematical systems using archaeology, historical analysis, and computer modeling. These investigations reveal that ancient mathematics was not primitive—it was creative, practical, and often surprisingly advanced.

The greatest lesson from ancient mathematics is that human curiosity has always searched for hidden patterns. The same desire that led ancient astronomers to track planets and ancient builders to measure stone also drives modern scientists exploring galaxies and quantum systems.

Numbers have always connected humanity to the universe.

Long before digital technology transformed the world, ancient civilizations were already asking some of the most important questions:

How does nature work? What patterns shape our existence? How can we measure the world around us?

Their answers were written in stone, clay, and manuscripts—but the ideas behind them remain alive.

The secret mathematics of the ancients was never truly lost. It became the foundation upon which future generations continued to build.

Chapter 10
The First Environmental Engineers

When people think of environmental engineering, they often imagine modern scientists designing renewable energy systems, restoring damaged ecosystems, or developing technologies to address climate challenges. Yet thousands of years ago, ancient civilizations were already solving environmental problems with remarkable creativity.

They built systems to manage water, protect soil, control floods, improve agriculture, and adapt to changing landscapes. Without modern machinery or advanced scientific instruments, they observed nature carefully and developed solutions that allowed communities to survive and thrive.

These early environmental engineers understood something that modern societies are still learning: human survival depends on maintaining a relationship with the natural world.

Ancient civilizations were not always perfectly sustainable, and some made decisions that contributed to environmental decline. However, many developed innovative approaches to working with their surroundings. Their achievements reveal a sophisticated understanding of ecosystems, resources, and adaptation.

One of humanity's earliest environmental challenges was water management. Reliable access to water determined where civilizations could develop. Rivers provided transportation, agriculture, and drinking water, but they also brought risks through flooding and unpredictable changes.

The people of ancient Mesopotamia were among the first to create large-scale water management systems. Living between the Tigris and Euphrates Rivers, they faced both opportunity and danger. Seasonal flooding could enrich farmland with fertile soil, but uncontrolled water could destroy crops and settlements.

To address these challenges, engineers built canals, levees, reservoirs, and irrigation networks. These systems redirected water to agricultural fields and allowed communities to produce food on a much larger scale.

Their innovations helped transform scattered settlements into some of the world's earliest cities.

Ancient Egyptian engineers developed a different relationship with the Nile River. Rather than attempting to completely control the river, they adapted their agricultural systems around its natural cycles. The annual flooding of the Nile deposited nutrient-rich soil across farmland, allowing farmers to grow crops in one of the world's most productive agricultural regions.

Egyptians created canals, basins, and irrigation methods that helped distribute water and maximize agricultural output. Their success depended on understanding the rhythm of the environment and working within natural patterns.

In South Asia, the Indus Valley Civilization demonstrated extraordinary urban environmental planning. Cities such as Mohenjo-daro and Harappa contained carefully designed drainage systems, wells, and water management structures.

Many homes had access to bathing areas connected to organized drainage networks. Streets were planned with sanitation in mind, and waste systems were incorporated into the design of the city itself. These achievements demonstrate that public health and environmental management were important priorities thousands of years ago.

The Indus cities reveal a powerful idea: a civilization's success depends not only on impressive buildings but also on the invisible systems that support daily life.

Ancient Chinese civilizations also developed environmental engineering solutions to manage water and agriculture. One of the most significant examples is the Dujiangyan irrigation system, created more than two thousand years ago in present-day China.

Unlike many flood-control projects that relied on massive barriers, Dujiangyan worked with the natural movement of the river. Engineers designed channels and structures that redirected water, reduced flooding, and provided irrigation without blocking the river completely.

The system continues to function today, demonstrating the effectiveness of designing solutions that cooperate with natural processes rather than fighting against them.

In the Andes Mountains, the Inca civilization created extraordinary agricultural engineering systems adapted to difficult terrain. Mountain environments presented challenges such as steep slopes, erosion, and unpredictable weather.

The Inca responded by building terraces carved into mountainsides. These terraces reduced soil erosion, improved drainage, and created microclimates that allowed crops to grow at different elevations. They transformed challenging landscapes into productive agricultural environments.

The Inca also developed sophisticated water channels and storage systems to support communities in high-altitude regions. Their engineering demonstrated a deep understanding of how geography and climate influence human survival.

Ancient communities also developed methods of soil management. Farmers around the world experimented with crop rotation, composting, irrigation techniques, and methods for preserving soil fertility.

In the Amazon region, researchers have discovered evidence that Indigenous communities created highly fertile soils known as terra preta, or "dark earth." These soils contain charcoal, organic material, and nutrients that improved agricultural productivity. Their creation suggests advanced knowledge of land management and ecological relationships.

Forests, too, were carefully managed by many Indigenous cultures. Rather than viewing natural environments as untouched wilderness, many communities actively shaped landscapes through controlled burning, selective planting, and resource management practices. These methods helped maintain biodiversity and support long-term survival.

Ancient environmental engineering was not limited to agriculture. Cities also required careful planning to manage heat, waste, and resources.

Some civilizations designed buildings to work with natural conditions. Structures were positioned to maximize shade, encourage airflow, and reduce exposure to extreme temperatures. In desert regions, architects developed methods for cooling buildings without modern electricity.

The ancient city of Petra, built by the Nabataeans, provides another remarkable example of environmental adaptation. Located in a dry desert environment, Petra survived because its inhabitants developed an advanced water collection system. They created channels, dams, and underground storage areas that captured rainfall and protected the city from flash floods.

Their engineering allowed a thriving urban center to exist in one of the world's harshest environments.

However, history also provides warnings. Environmental engineering could not always prevent collapse. Some civilizations overused resources, damaged ecosystems, or failed to adapt to changing conditions. Deforestation, soil depletion, and water mismanagement contributed to difficulties faced by several ancient societies.

These examples reveal an important lesson: technology alone cannot guarantee survival. The success of a civilization depends on how wisely it uses resources and responds to environmental changes.

Modern researchers continue studying ancient environmental systems because many of their ideas remain relevant. Sustainable agriculture, water conservation, natural cooling methods, and ecosystem-based planning are increasingly important as societies face modern environmental challenges.

Ancient engineers understood principles that remain central today: water must be managed carefully, soil must be protected, and human communities must respect the limits of their environment.

The first environmental engineers did not have advanced laboratories or computer models. Their tools were observation, experience, experimentation, and cooperation with nature. They learned by watching rivers, studying seasons, and understanding the landscapes that surrounded them.

Their achievements remind us that environmental innovation is not a new idea. Humanity has been searching for ways to live sustainably for thousands of years.

The solutions created by ancient civilizations were not perfect, but they reveal something remarkable about human creativity. Long before the modern environmental movement existed, people were already designing systems to protect resources, adapt to challenges, and build a future for generations to come.

The first environmental engineers were not defined by the technology they possessed. They were defined by their ability to understand the world around them—and their determination to live within it.

Chapter 11
Maps That Predicted the Unknown

A map is more than a drawing of places. It is a record of human curiosity, exploration, imagination, and ambition. Every line, symbol, and marking represents someone attempting to understand the world beyond what they could immediately see.

For thousands of years, maps guided travelers across oceans, deserts, and mountains. They helped civilizations trade, expand, navigate, and communicate. Yet some ancient and historical maps contain details that appear far ahead of their time, raising fascinating questions about how earlier societies gathered information about the world around them.

While many legendary maps have been surrounded by exaggerated claims and speculation, genuine historical discoveries reveal something equally remarkable: ancient cartographers possessed extraordinary observational skills, preserved knowledge from earlier explorers, and sometimes understood geography far better than historians once believed.

The creation of maps required more than artistic ability. It demanded mathematics, astronomy, surveying, navigation, and an understanding of landscapes. Before satellites and modern instruments existed, mapmakers relied on careful measurements, travelers' accounts, celestial observations, and generations of accumulated knowledge.

One of the earliest known examples of mapping comes from ancient Mesopotamia. Clay tablets created thousands of years ago contained representations of cities, regions, and land boundaries. These early maps were not designed like modern geographic charts, but they demonstrated humanity's desire to organize and record space.

The Babylonians created some of the earliest known world maps, including the famous Babylonian Map of the World. Although it reflected ancient beliefs about geography, it also provided insight into how people of the time understood their environment. It showed cities, rivers, and surrounding regions while blending practical information with cultural perspectives.

Ancient Egyptian civilization also developed advanced mapping techniques. Surveyors used measurements to define land boundaries, especially after the annual flooding of the Nile changed property lines. Their ability to restore accurate boundaries contributed to the development of geometry and practical mathematics.

Egyptian maps also supported exploration. Records from expeditions into distant regions demonstrate that ancient travelers collected geographic information about foreign lands, resources, and routes.

The Greeks transformed mapmaking by introducing mathematical geography. Instead of simply representing locations, Greek scholars attempted to calculate the size and shape of the Earth.

The mathematician and geographer Eratosthenes made one of the most impressive calculations of the ancient world. Using observations of shadows in different locations, he estimated Earth's circumference with remarkable accuracy. His work demonstrated that geography could be based on measurement rather than assumption.

Greek geographers also developed early systems of latitude and longitude, concepts that became essential for modern navigation. Their methods allowed future generations to create increasingly accurate representations of the planet.

One of history's most fascinating maps is the Ptolemaic world map, based on the work of the ancient Greek scholar Claudius Ptolemy. Created in the second century CE, it included thousands of locations and a coordinate system for organizing geographic information.

Although many details were inaccurate because of limited knowledge of distant regions, Ptolemy's approach influenced mapmaking for more than a thousand years. His work demonstrated the importance of mathematical methods in understanding the world.

During the medieval period, knowledge from different civilizations continued to shape cartography. Islamic scholars preserved and expanded earlier geographic traditions while creating their own maps based on exploration, astronomy, and trade knowledge.

The work of scholars such as Al-Idrisi demonstrated the sophistication of medieval geography. In the twelfth century, Al-Idrisi created a detailed world map and geographic encyclopedia for the Norman court of Sicily. His work combined information from travelers, merchants, and earlier sources, creating one of the most advanced geographic references of the medieval era.

Trade networks played a major role in improving maps. Merchants traveled through unfamiliar territories and recorded information about coastlines, cities, rivers, and routes. Their practical knowledge became valuable to future explorers.

One of the most intriguing historical maps is the Piri Reis map, created in the early sixteenth century by Ottoman admiral and cartographer Piri Reis. The surviving portion depicts parts of the Atlantic Ocean, including sections of South America and nearby islands.

The map has inspired many debates because of its unusual details and the sources Piri Reis claimed to have used, including earlier charts and information from explorers. Historians generally view it as an example of how cartographers combined multiple sources rather than evidence of impossible knowledge. Its importance lies in showing how geographic information was collected, compared, and preserved across cultures.

Another fascinating example is the Tabula Peutingeriana, a medieval copy of an ancient Roman road map. Unlike modern maps designed for accurate geographic scale, this map emphasized routes and travel connections. It functioned more like an ancient travel guide, showing roads, settlements, and distances across the Roman world.

Maps have also revealed knowledge of places before they became widely known. Viking explorers created navigational traditions that allowed them to cross the North Atlantic centuries before many Europeans understood the full geography of the region. Their journeys demonstrated remarkable knowledge of ocean currents, stars, weather patterns, and coastal navigation.

Indigenous mapping traditions provide another important perspective. Many cultures created maps using materials and methods different from European traditions. Polynesian navigators used sophisticated systems based on ocean swells, winds, stars, and animal behavior to navigate thousands of miles across the Pacific.

Their knowledge was not always recorded on paper, but it was preserved through oral traditions, training, and practical experience. These navigation systems demonstrate that maps can exist as knowledge passed from person to person, not only as drawings.

The history of cartography also reveals how much of the world remained unknown for centuries. Early maps often contained empty spaces labeled with uncertainty. These areas represented not ignorance, but opportunity—the places where future explorers would gather new information.

As exploration expanded, maps changed. Coastlines became more accurate. New continents appeared. Previously unknown islands were documented. The blank spaces slowly disappeared.

Today, satellites and digital technology have created maps of extraordinary precision. Yet modern mapping still depends on principles developed by ancient observers: measurement, comparison, calculation, and curiosity.

Perhaps the most fascinating aspect of historical maps is not the places they accurately predicted, but the questions they inspired. A map represents a moment in time when humanity understood part of the world but still searched for what remained hidden.

Every map is both a record and an invitation. It tells us where we have been while encouraging us to explore what remains unknown.

The ancient cartographers who studied stars, measured landscapes, and recorded distant journeys were not simply drawing locations. They were creating a bridge between imagination and discovery.

Their maps remind us that humanity has always looked beyond the visible horizon. The unknown has never been a barrier—it has been a reason to explore.

And sometimes, the maps created by those who came before us reveal that the world was already being discovered long before we realized it.

Chapter 12
Civilizations Built Around Water

Water has always been more than a resource. It has been the foundation upon which civilizations were built, the force that shaped economies, determined settlements, influenced cultures, and guided the rise and fall of societies.

From the earliest river valleys to magnificent coastal cities, humans have always gathered near water. Rivers provided transportation, fertile soil, food sources, and opportunities for trade. Oceans connected distant communities and encouraged exploration. Lakes and underground water systems allowed people to survive in challenging environments.

Yet water was both a gift and a challenge. Civilizations that learned to manage it flourished. Those that failed to adapt to changing water conditions often faced decline.

The history of humanity is, in many ways, a history of water.

One of the earliest examples of a civilization shaped by water was ancient Mesopotamia, located between the Tigris and Euphrates Rivers. Often called the "cradle of civilization," this region became home to some of the world's first cities, governments, writing systems, and organized societies.

The rivers provided fertile soil that allowed agriculture to expand. However, unlike the predictable flooding of the Nile, the rivers of Mesopotamia could be unpredictable and destructive. Communities developed complex irrigation systems to control water flow and support farming.

Canals, reservoirs, and drainage networks transformed dry landscapes into productive farmland. These engineering achievements allowed cities such as Uruk, Ur, and Babylon to grow into powerful centers of culture and commerce.

The success of Mesopotamia depended on cooperation. Managing water required large-scale planning, labor organization, and shared responsibility. In many ways, the need to control rivers helped encourage the development of governments and administrative systems.

Ancient Egypt provides another remarkable example of a civilization shaped by water. The Nile River was the center of Egyptian life. Each year, seasonal flooding deposited nutrient-rich soil along the riverbanks, creating some of the most fertile agricultural land in the ancient world.

The Egyptians developed their calendar around the Nile's cycles. Farming, transportation, religious practices, and economic systems were all connected to the river.

The Nile was often described as the source of life itself. It allowed a civilization to thrive in a landscape surrounded by desert. Without the river, the great temples, cities, and monuments of ancient Egypt could never have existed.

Across the Atlantic, the Maya civilization developed sophisticated water systems in the tropical environment of Mesoamerica. Unlike Egypt and Mesopotamia, many Maya cities were not located beside massive rivers. Instead, they depended on rainfall collection, reservoirs, underground water sources, and carefully designed storage systems.

Large Maya cities created reservoirs to capture and preserve seasonal rainfall. These systems supported growing populations during dry periods and demonstrated advanced environmental planning.

Water management was closely connected to Maya architecture and urban design. Their cities were carefully planned around available resources, showing that survival depended on understanding local environments.

In the Andes Mountains, the Inca civilization faced a different challenge. High elevations, steep terrain, and unpredictable weather made water management essential.

The Inca built channels, fountains, agricultural terraces, and reservoirs that allowed communities to survive in difficult mountain environments. Their terraces slowed water runoff, reduced erosion, and helped create productive farmland.

The famous city of Machu Picchu demonstrates this relationship between civilization and water. Built high in the mountains, the city included carefully engineered drainage systems that protected buildings from heavy rainfall. Its survival for centuries is a testament to the skill of its builders.

Water also shaped civilizations through maritime trade. Coastal societies developed powerful economies by using oceans as highways connecting distant regions.

The Phoenicians, who lived along the eastern Mediterranean coast, became legendary sailors and traders. Their ships traveled across the Mediterranean Sea carrying goods such as timber, glass, metals, and textiles. Their maritime expertise helped spread ideas, technologies, and cultural influences throughout the region.

Ancient Greece was similarly shaped by the sea. Its mountainous landscape made farming difficult in many areas, but its extensive coastline encouraged naval exploration and trade. The sea influenced Greek politics, warfare, mythology, and identity.

In Southeast Asia, water-based civilizations developed around rivers, deltas, and coastal environments. The Khmer Empire created the extraordinary city of Angkor, supported by one of the largest water management systems of the ancient world.

The Khmer built reservoirs, canals, and embankments that controlled seasonal flooding and supported intensive agriculture. Their ability to manage water helped sustain a massive urban population for centuries.

However, water could also become a source of vulnerability. Environmental changes, droughts, floods, and shifts in waterways contributed to the decline of several civilizations.

Angkor's decline, for example, has been linked in part to changing climate patterns that placed stress on its complex water infrastructure. Researchers believe cycles of intense drought followed by heavy flooding may have damaged the system beyond its ability to recover.

This pattern appears repeatedly throughout history. Civilizations often depend on delicate relationships with their environments. The same resource that creates prosperity can become a challenge when conditions change.

Water also preserved history in unexpected ways. Flooded settlements, submerged ports, and underwater ruins have revealed information about civilizations that disappeared beneath lakes, rivers, and oceans.

Archaeologists have discovered ancient shipwrecks containing trade goods, tools, and personal belongings that provide glimpses into past societies. Submerged cities reveal that human communities existed in places now transformed by rising waters or geological changes.

Modern technology has expanded the search for these hidden worlds. Underwater archaeology, satellite imaging, and advanced mapping techniques allow researchers to explore ancient landscapes that were once inaccessible.

The lessons from water-based civilizations remain relevant today. Modern societies continue to face challenges involving water access, flooding, drought, and environmental change. Ancient solutions do not provide perfect answers, but they demonstrate the importance of planning, adaptation, and respect for natural systems.

The greatest water engineers of the past understood something fundamental: humans cannot control nature completely. They must learn how to work with it.

Rivers must be guided, not defeated. Rain must be collected, not wasted. Oceans must be understood, not feared.

The civilizations that thrived around water succeeded because they recognized its power. They built cities where rivers flowed, developed technologies to manage resources, and created cultures deeply connected to their environments.

Their achievements remind us that water is not simply something civilizations use—it is something civilizations are built around.

From ancient river valleys to modern coastal cities, humanity's story has always followed the movement of water. It is the thread that connects the earliest settlements to the world we know today.

Where water flowed, civilizations followed. Where people learned to understand it, civilizations flourished.

Chapter 13
Lost Agricultural Breakthroughs

Before cities rose, before empires expanded, and before written history began, humanity's greatest challenge was simple: finding enough food to survive.

Agriculture transformed the human story. It allowed communities to settle in one place, develop specialized skills, build cities, create governments, and preserve knowledge across generations. Yet the history of farming is not only a story of progress. It is also a story of forgotten discoveries, abandoned techniques, and agricultural innovations that disappeared as civilizations changed.

Many ancient farming methods were remarkably advanced. They reflected generations of observation and experimentation with soil, climate, water, plants, and ecosystems. Some were abandoned because of political collapse, environmental changes, or cultural disruption. Others were replaced by newer methods before their full value was understood.

Today, scientists and farmers are rediscovering these lost agricultural breakthroughs and examining how ancient knowledge may help address modern challenges such as climate change, soil degradation, and food security.

The earliest farmers were innovators who transformed wild landscapes into productive environments. The domestication of plants and animals was not a single discovery but a long process of experimentation. Early communities selected seeds from stronger plants, learned which crops grew best in specific conditions, and developed methods for improving harvests.

Over thousands of years, farmers around the world created systems uniquely adapted to their environments. These methods were not random traditions—they were carefully developed solutions based on generations of experience.

One of the most remarkable examples comes from the ancient civilizations of the Andes Mountains. The Inca and earlier Andean cultures developed agricultural techniques that allowed communities to grow food in one of the world's most challenging landscapes.

Mountain environments presented serious obstacles. Steep slopes caused erosion, temperatures changed dramatically with elevation, and rainfall patterns were unpredictable. Ancient farmers responded by creating agricultural terraces carved into mountainsides.

These terraces did more than provide flat land for planting. They reduced erosion, improved water retention, created different growing conditions at various elevations, and protected crops from extreme weather. Some terraces produced multiple microclimates within a small area, allowing farmers to grow a wider variety of crops.

Many abandoned terraces are now being restored because researchers recognize their potential for sustainable farming in mountainous regions.

Another forgotten agricultural achievement was the creation of terra preta, or "dark earth," in the Amazon region. For many years, researchers struggled to understand how fertile soils existed in areas where tropical environments typically contain nutrient-poor ground.

Studies revealed that Indigenous communities created these highly productive soils by adding charcoal, organic waste, plant materials, and other nutrients over long periods.

Unlike many modern agricultural practices that can exhaust soil, terra preta improved fertility and remained productive for centuries. Scientists continue studying these ancient methods to better understand long-term soil restoration.

The Maya civilization also developed impressive agricultural innovations. Living in tropical environments with seasonal rainfall, Maya farmers created systems to maximize food production despite environmental challenges.

They built raised fields, managed forests, collected rainwater, and developed diverse farming systems. Rather than relying on a single crop, they cultivated multiple plants that supported nutrition and reduced the risk of complete agricultural failure.

Modern research has challenged older assumptions that ancient tropical civilizations simply cleared forests and exhausted their environments. Evidence suggests many Indigenous farming practices involved careful landscape management and ecological knowledge.

Ancient Chinese agricultural methods also contained valuable innovations. Farmers developed techniques for maintaining soil fertility, managing water, and increasing crop production over long periods.

One important practice was the use of organic fertilizers and crop rotation. By understanding that soil required replenishment, ancient farmers developed systems that prevented exhaustion and supported continued productivity.

In parts of Asia, rice cultivation became an extraordinary example of environmental adaptation. Farmers created intricate irrigation systems and carefully managed water levels to support rice growth. These systems required cooperation, planning, and detailed knowledge of local landscapes.

Ancient Egypt's agricultural success also depended on innovation. Farmers used the predictable flooding of the Nile River to enrich farmland naturally. They developed irrigation methods, planting schedules, and agricultural tools designed specifically for their environment.

The Egyptian approach demonstrates a recurring theme in ancient agriculture: successful farming came from understanding natural cycles rather than attempting to completely overcome them.

Some ancient agricultural breakthroughs involved plants themselves. Farmers selectively bred crops for thousands of years, creating varieties better suited for different climates and purposes.

Modern foods such as corn, wheat, rice, and many fruits and vegetables are the result of generations of human experimentation. Ancient farmers were early genetic innovators, shaping plant characteristics long before modern genetics existed.

They selected plants with larger seeds, better resistance, improved taste, or higher productivity. Each generation built upon the choices of those before them.

Ancient farmers also developed sophisticated methods of water conservation. In dry regions, communities created underground channels, reservoirs, and collection systems that allowed agriculture to continue despite limited rainfall.

The qanat system, developed in ancient Persia, provided a remarkable example of underground water management. These gently sloping tunnels transported groundwater from elevated areas to settlements and fields while reducing evaporation. This technology allowed communities to farm in some of the world's driest environments.

The disappearance of agricultural knowledge often occurred during periods of social disruption. When civilizations collapsed, farming systems that required specialized knowledge could be abandoned. Political changes, population movement, and cultural replacement sometimes caused valuable techniques to disappear.

However, agricultural knowledge rarely vanishes completely. It often survives through traditional practices maintained by communities, even when its original history is forgotten.

Today, scientists are returning to ancient agricultural methods with new tools. Soil analysis, environmental research, and climate modeling allow researchers to understand why certain traditional systems worked so effectively.

Ancient techniques are being studied for their potential contributions to modern problems. Improved soil management, water conservation, biodiversity protection, and climate-resilient farming are areas where forgotten knowledge may offer valuable lessons.

This does not mean returning entirely to ancient farming methods. Modern agriculture has produced extraordinary achievements in productivity and food availability. Instead, the goal is to combine the strengths of modern science with the wisdom gained through thousands of years of human experimentation.

The story of agriculture is ultimately a story of adaptation. Farmers throughout history faced changing climates, limited resources, and unpredictable conditions. They survived because they learned to observe, experiment, and innovate.

Many of their discoveries were never written in books or preserved in laboratories. They existed in fields, passed from one generation to the next through practice and experience.

The lost agricultural breakthroughs of the past remind us that innovation does not always appear in a laboratory or a factory. Sometimes it grows quietly in the soil beneath our feet.

The farmers of ancient civilizations were scientists in their own way. They studied nature, tested ideas, and created solutions that allowed humanity to flourish.

Their discoveries may have been forgotten for a time, but the land still remembers what they knew.

Chapter 14

The Forgotten Women Who Changed History

History is often written around the names of kings, conquerors, inventors, and political leaders. Their achievements fill textbooks, monuments, and museums. Yet behind many of the world's greatest advancements are women whose contributions were overlooked, minimized, or forgotten entirely.

For centuries, women faced barriers that limited their access to education, professional recognition, political influence, and ownership of their own discoveries. Despite these challenges, countless women shaped science, medicine, mathematics, literature, exploration, technology, and social progress.

Some worked in royal courts. Others taught in universities, studied the natural world, created inventions, preserved knowledge, or challenged the boundaries of what society believed women could accomplish.

Their stories reveal a powerful truth: progress has never belonged to one group of people. Civilization has been built by countless minds, including many whose names were nearly lost to time.

One of the earliest known female scholars was Enheduanna of ancient Mesopotamia. Living more than four thousand years ago, she was a high priestess, poet, and administrator in the Sumerian city of Ur. She is often considered one of the earliest known authors in history whose name can be connected to surviving written works.

Her hymns and writings provide insight into ancient religion, politics, and literature. At a time when most historical records focused on rulers and warfare, Enheduanna's words preserved the voice of an individual woman who influenced culture and intellectual life.

Ancient Egypt also produced remarkable women who contributed to leadership, science, and knowledge. Figures such as Hatshepsut demonstrated that women could hold positions of immense political power. As one of Egypt's most successful rulers, she expanded trade networks, commissioned architectural projects, and strengthened the prosperity of her kingdom.

Her achievements were significant enough that later attempts were made to remove evidence of her reign. Yet archaeology eventually restored her place in history, showing how powerful figures could be deliberately forgotten.

In ancient Alexandria, the mathematician and philosopher Hypatia became one of the most respected intellectual figures of her era. She taught mathematics, astronomy, and philosophy during a period when women rarely held public scholarly roles.

Although much of her original writing has been lost, historical accounts describe her as a brilliant teacher and thinker who contributed to the preservation and advancement of scientific knowledge. Her life represents the importance of protecting intellectual traditions, especially during times of social and political conflict.

Women also played crucial roles in the development of medicine. In many ancient and medieval societies, women served as healers, midwives, and caretakers who preserved practical medical knowledge.

Much of this knowledge was passed through generations rather than recorded in formal institutions. As a result, many women's contributions disappeared from written history even though their skills were essential to community health.

During the medieval period, women continued to influence scholarship and innovation. In European monasteries, women copied manuscripts, studied natural remedies, and preserved ancient texts. These efforts helped maintain knowledge during a time when written works were vulnerable to destruction.

One remarkable figure was Hildegard of Bingen, a twelfth-century scholar, writer, composer, and naturalist. She wrote extensively about medicine, plants, philosophy, and spirituality. Her observations of nature and health reflected an unusually broad understanding of the world around her.

Her writings demonstrate that intellectual achievement was not limited to traditional academic institutions.

The history of science also contains many examples of women whose work transformed modern understanding. Caroline Herschel, an eighteenth-century astronomer, discovered several comets and contributed important observations to astronomy. Although her brother William Herschel is often more widely remembered, Caroline's discoveries earned her recognition as one of the important astronomers of her time.

Mathematics and computing also owe much to overlooked women. Ada Lovelace, often considered one of the first computer programmers, recognized the potential of Charles Babbage's mechanical computing designs in the nineteenth century. Her notes described how machines could potentially do more than simple calculations, anticipating concepts that would become central to modern computing.

Her ideas were far ahead of the technology available during her lifetime, and their importance became clearer as computer science developed.

Women also shaped scientific research through observation and discovery. Botanists, chemists, physicians, and researchers contributed to fields that were often dominated by men. Many faced challenges receiving credit for their work, yet their discoveries continued to influence future generations.

A powerful example is the role of women in understanding DNA. Scientists such as Rosalind Franklin produced critical research images that contributed to the discovery of DNA's structure. Her contributions were not fully recognized during her lifetime, but later historical assessments highlighted the importance of her work.

Women also changed history outside of science. Writers, activists, explorers, artists, and reformers challenged social expectations and expanded possibilities for future generations.

The explorer Jeanne Baret became the first woman known to have circumnavigated the globe. Disguised as a man because women were prohibited from joining certain expeditions, she participated in scientific exploration and contributed botanical knowledge during an era when women had limited opportunities in the sciences.

Across cultures, women preserved traditions, languages, stories, and knowledge that might otherwise have disappeared. They served as educators, community leaders, innovators, and creators.

Yet the question remains: Why were so many women's achievements forgotten?

The answer is complex. Historical records were often created by societies where men held greater political and educational power. Women frequently lacked access to institutions that preserved knowledge. Their work was sometimes credited to male colleagues, dismissed as informal labor, or excluded from official accounts.

However, modern historians, archaeologists, and researchers continue uncovering these overlooked stories. New evidence, restored documents, and revised interpretations are revealing a more complete picture of human achievement.

Recognizing forgotten women does not diminish the accomplishments of those already celebrated. Instead, it creates a more accurate understanding of history. Innovation has always been a collective effort, shaped by people from many backgrounds and circumstances.

The women featured in history's overlooked chapters share a common trait: they pursued knowledge despite obstacles. They questioned assumptions, solved problems, created new ideas, and contributed to a world that did not always acknowledge them.

Their stories remind us that greatness is not always accompanied by fame. Some of the most important contributions in human history came from people whose names were never carved into monuments.

The forgotten women who changed history are not forgotten anymore.

Their discoveries, writings, inventions, and courage continue to influence the modern world. By bringing their stories back into view, we gain a fuller understanding of civilization itself—not as the achievement of a few famous individuals, but as the result of millions of voices working together across time.

History was never created by only those whose names survived.

It was shaped by everyone who dared to imagine a different future.

Chapter 15
Invisible Empires of Commerce

When people think of ancient empires, they often imagine armies, palaces, monuments, and powerful rulers. They picture kingdoms that expanded through conquest and maintained control through military strength. Yet some of history's most influential powers were not built primarily through warfare.

They were built through commerce.

Across the ancient world, trading networks created invisible empires—systems of exchange that connected distant regions, shaped economies, influenced cultures, and transformed societies. These commercial powers did not always leave behind massive monuments or famous royal dynasties, but their influence reached across continents.

Their strength came from relationships, knowledge, transportation routes, financial systems, and the ability to connect people who lived thousands of miles apart.

Commerce was one of humanity's earliest forms of globalization. Long before modern corporations and international markets existed, merchants created networks that linked civilizations through the exchange of goods, ideas, and technology.

These invisible empires demonstrate that power has never belonged only to those who controlled land. Sometimes, the people who controlled connections controlled history.

One of the earliest examples of a commercial civilization was the Phoenician network of the ancient Mediterranean. The Phoenicians lived along the eastern Mediterranean coastline in city-states such as Tyre, Sidon, and Byblos. Their homeland contained limited agricultural land, encouraging them to turn toward maritime trade.

They became exceptional sailors, shipbuilders, and merchants. Their ships traveled throughout the Mediterranean, establishing trading settlements and connecting distant communities.

The Phoenicians traded valuable goods such as timber, glass, textiles, metals, and the famous purple dye known as Tyrian purple. This dye became associated with royalty because of its rarity and expensive production process.

Their greatest influence, however, may have been their alphabet. As merchants interacted with different cultures, they used a practical writing system that was easier to learn than many earlier scripts. Variations of the Phoenician alphabet influenced later writing systems, including those used by the Greeks and Romans.

Their commercial empire was not defined by conquered territories but by the relationships they created across the sea.

Another powerful commercial network emerged through the ancient Silk Road. Although often described as a trade route, it functioned more like a massive economic system connecting merchants, cities, and civilizations across Eurasia.

No single empire controlled the entire Silk Road. Instead, different groups benefited from controlling key locations, protecting routes, and facilitating exchanges.

Merchants transported silk, spices, precious metals, horses, ceramics, glass, and luxury goods across enormous distances. However, the movement of information may have been even more valuable than the movement of products.

Ideas about mathematics, medicine, astronomy, religion, agriculture, and technology traveled alongside merchants. A marketplace could become a meeting point where civilizations exchanged not only goods but knowledge.

In ancient Africa, powerful trading networks created prosperous societies that connected the interior of the continent with international markets.

The kingdoms of West Africa became wealthy through their control of valuable resources, especially gold and salt. Gold from regions south of the Sahara traveled through desert trade routes toward North Africa and beyond.

The Ghana Empire, followed by Mali and Songhai, developed economies supported by commerce. Their wealth allowed them to build cities, support scholarship, and establish political influence.

The city of Timbuktu became a famous center of trade and learning. Merchants gathered there alongside scholars, exchanging goods and ideas from across Africa and the wider world.

These examples challenge outdated views that ancient Africa was isolated. In reality, African civilizations participated in extensive international networks that connected them to Europe, the Middle East, and Asia.

Across the Indian Ocean, another invisible empire of commerce developed through maritime trade. For centuries, merchants from Arabia, India, East Africa, Southeast Asia, and China used seasonal monsoon winds to navigate predictable ocean routes.

Unlike land-based empires that depended on controlling territory, these trading networks depended on ports, trust, contracts, and cultural connections.

Major port cities became international crossroads where merchants from different backgrounds lived and worked together. Goods such as spices, textiles, ceramics, ivory, and precious stones moved across the ocean, creating economic relationships that lasted for generations.

The power of these commercial networks came from cooperation. A merchant could travel thousands of miles because they relied on established customs, agreements, languages, and systems of exchange.

Ancient commerce also required sophisticated financial innovations. Merchants developed methods for managing risk, recording transactions, extending credit, and transferring value over long distances.

In Mesopotamia, written records on clay tablets documented business agreements, loans, and trade transactions. These records reveal that ancient economies were far more complex than simple barter systems.

Merchants needed systems of trust because long-distance trade involved uncertainty. A trader might invest in goods that would not reach their destination for months or years. Contracts, partnerships, and reputation became essential parts of commercial success.

The ancient world also saw the rise of specialized merchant communities. These groups often maintained networks that extended across political boundaries. Their knowledge of languages, customs, and markets allowed them to operate successfully in diverse environments.

In some ways, these merchant networks functioned like early multinational organizations. They connected distant economies, adapted to changing political conditions, and created influence without directly ruling large territories.

However, invisible empires of commerce were also vulnerable. Trade depended on stability. Wars, piracy, changing political borders, natural disasters, and environmental shifts could disrupt even the strongest networks.

When major trade routes declined, cities that depended on commerce often suffered. Ports became quieter, markets disappeared, and once-thriving communities declined.

Yet the impact of these commercial systems remained. Even after trade networks changed, the ideas, technologies, and cultural exchanges they created continued shaping the world.

Modern economies still reflect many principles developed by ancient merchants. Global supply chains, international markets, financial systems, and cultural exchanges all have roots in these early networks of connection.

The invisible empires of commerce reveal an important lesson about power. Influence does not always come from armies or territory. Sometimes it comes from the ability to connect people, move resources, and create relationships across boundaries.

The merchants of the ancient world were not merely sellers of goods. They were explorers, negotiators, translators, and carriers of civilization's shared knowledge.

Their ships and caravans created pathways between worlds that once seemed separate.

Although many of their names have disappeared, their impact remains everywhere. The foods we eat, the languages we speak, the technologies we use, and the cultures we experience have all been shaped by centuries of exchange.

The greatest empires were not always the ones that built the largest monuments.

Some of the most powerful were the ones that quietly connected the world.

Chapter 16
Technologies That Disappeared

Human history is often described as a steady march of progress—a journey from simple tools to advanced machines, from ancient inventions to modern technology. However, this view overlooks a fascinating reality: progress is not always a straight line.

Throughout history, civilizations have created remarkable technologies that were later forgotten, abandoned, or lost. Some disappeared because societies collapsed. Others vanished because materials were destroyed, knowledge was not passed down, or newer methods replaced older solutions. In some cases, entire technologies remained hidden for centuries until archaeologists uncovered evidence of their existence.

These lost technologies challenge the idea that the past was always less advanced than the present. They reveal that earlier civilizations possessed creativity, engineering skill, and scientific understanding that sometimes rivaled later achievements.

The story of forgotten technology is not only about machines. It is about human knowledge—how it is created, preserved, shared, and sometimes lost.

One of the most extraordinary examples of ancient technology is the Antikythera mechanism, discovered in a shipwreck near the Greek island of Antikythera in 1901. At first, the corroded bronze fragments appeared to be an ordinary artifact. However, detailed analysis revealed something astonishing: a complex mechanical device containing gears designed to calculate astronomical cycles.

Created more than two thousand years ago, the mechanism demonstrated a level of mechanical sophistication that surprised researchers. It could track the movements of celestial bodies and predict astronomical events using carefully designed gear systems.

The exact creator remains unknown, and the knowledge required to build such a device appears to have disappeared for centuries. The Antikythera mechanism serves as a reminder that advanced ideas can emerge long before societies develop the ability to preserve them.

Ancient Rome also possessed technologies that were later lost or forgotten. Roman engineers developed extraordinary construction methods, including durable concrete that allowed structures such as harbors, bridges, and buildings to survive for centuries.

After the fall of the Roman Empire, many specialized building techniques declined. The knowledge of certain materials, formulas, and construction methods was not preserved in the same way, causing later builders to spend centuries rediscovering techniques that ancient engineers had already mastered.

The Roman approach to infrastructure was equally impressive. Their roads, aqueducts, heating systems, and sanitation networks demonstrated advanced planning and engineering. Some technologies, such as underfloor heating systems known as hypocausts, disappeared from widespread use for centuries before being replaced by newer heating methods.

Another fascinating example comes from ancient Greece: automated mechanical devices. Engineers such as Hero of Alexandria designed machines powered by air pressure, steam, and mechanical principles. Some of these devices included automatic doors, theatrical machines, and mechanisms that created movement without direct human control.

Although these inventions were not used to create an industrial revolution during their time, they demonstrated a remarkable understanding of mechanical engineering. Many concepts appeared again centuries later in different forms.

Ancient China produced numerous technologies that transformed human history but were later forgotten or developed differently in other regions. Chinese inventors created innovations such as advanced agricultural tools, mechanical devices, printing techniques, and early forms of navigation technology.

One example is the south-pointing chariot, a mechanical device designed to maintain a consistent directional reference without relying on a magnetic compass. Historical records describe such devices, though their exact construction methods remain uncertain.

Chinese engineers also developed sophisticated water-powered mechanisms, including astronomical clocks and mechanical systems that demonstrated advanced understanding of gears, hydraulics, and automation.

In the Islamic Golden Age, scholars preserved and expanded knowledge from earlier civilizations while creating new technologies of their own. Engineers designed complex machines, including water-raising devices, automated instruments, and mechanical clocks.

Some of these inventions were documented in manuscripts, but many practical designs disappeared when political centers declined and institutions that supported scientific work were disrupted.

The loss of technology often occurred because knowledge depended on specialists. A technique might survive only as long as a small group of craftsmen, engineers, or scholars continued teaching it. If that tradition ended, the knowledge could vanish.

This happened with many ancient crafts. Specialized methods for producing metals, pigments, glass, textiles, and ceramics sometimes disappeared when communities migrated or trade networks collapsed.

The famous Roman purple dye known as Tyrian purple is an example of a lost production technique. Created from marine organisms through an extremely complex process, the dye became associated with wealth and power. After the ancient industries that produced it declined, the exact methods were largely forgotten.

Another remarkable example is ancient Damascus steel. Known historically for its strength and distinctive patterns, this material was produced using techniques that remain debated among researchers. The original manufacturing methods were eventually lost, leaving modern scientists to study surviving examples in an attempt to understand the process.

Some lost technologies were not forgotten because they failed. They disappeared because circumstances changed.

A society may abandon a technology because resources become unavailable, economic systems shift, or new methods become more practical. A successful invention in one environment may become unnecessary in another.

The disappearance of technology also reveals the importance of preserving knowledge. Ancient libraries, workshops, and educational institutions served as places where skills were recorded and passed forward. When these systems were damaged or destroyed, valuable information disappeared with them.

Today, researchers use archaeology, experimental reconstruction, and advanced analysis to recover lost technologies. Scientists recreate ancient materials, engineers rebuild historical machines, and historians examine old texts to understand how forgotten inventions worked.

Modern technology has also created new ways to preserve knowledge. Digital archives, three-dimensional scanning, and scientific databases allow information to be stored and shared globally.

Yet even modern society faces the same challenge as ancient civilizations: knowledge must be protected and passed on. A technology that is not understood or maintained can disappear regardless of how advanced it once was.

The forgotten technologies of history reveal that innovation is not limited to the modern era. Ancient inventors created solutions that reflected deep intelligence, creativity, and practical understanding.

Their achievements remind us that the past was not simply a collection of primitive attempts leading toward the present. It was filled with moments of brilliance—some celebrated, some hidden, and some waiting to be rediscovered.

Every lost technology tells a story about human imagination. It shows what people were capable of creating when they combined curiosity with determination.

The greatest lesson may be this: technology is not truly lost when its physical form disappears. As long as humanity continues asking questions, exploring evidence, and seeking understanding, forgotten inventions can return.

The past still holds machines we have not fully understood, ideas we have not recovered, and discoveries that may one day inspire the future.

Chapter 17
The Mystery of Impossible Construction

Throughout history, humans have created structures so extraordinary that they seem to challenge the limits of their time. Massive monuments, perfectly fitted stone walls, towering temples, and precisely engineered cities continue to inspire wonder thousands of years after their creation.

When modern observers encounter these achievements, one question often appears: **How did they build this?**

The answer is rarely magic or mystery. Instead, ancient construction reveals something far more impressive—the ability of humans to solve enormous problems through organization, experimentation, mathematics, craftsmanship, and determination.

Many ancient structures appear impossible only because we underestimate the knowledge and abilities of the people who created them. Their tools may have been different from ours, but their intelligence and creativity were just as powerful.

Across the world, ancient builders faced challenges that would test even modern engineers. They had to move enormous stones, design stable foundations, manage large labor forces, transport materials across difficult landscapes, and create structures capable of surviving earthquakes, floods, and centuries of weather.

Their accomplishments reveal forgotten chapters of human engineering.

Perhaps the most famous example is the Great Pyramid of Giza in Egypt. Built more than four thousand years ago, the pyramid remains one of the most remarkable construction achievements in history.

The structure contains millions of stone blocks, many weighing several tons. Its enormous scale has inspired countless theories about how ancient Egyptians transported and positioned the materials.

Archaeological evidence suggests that the pyramid was built through careful planning, skilled labor, and advanced organization. Workers used sledges, ramps, levers, surveying techniques, and coordinated teams to move materials. The construction process likely involved thousands of skilled workers supported by an organized society capable of managing food supplies, transportation, and logistics.

The true mystery is not that ancient people could build such a structure. The mystery is how effectively they organized such a massive project without modern machinery.

Another example of extraordinary construction can be found at Göbekli Tepe in modern-day Turkey. Dating back more than ten thousand years, this site challenged previous assumptions about early human societies.

The massive carved stone pillars, some weighing many tons, were created by communities that existed before traditional agriculture-based civilizations developed. The site suggests that early humans were capable of large-scale cooperation and monumental construction much earlier than previously believed.

Its discovery changed how researchers think about the development of complex societies. It demonstrated that monumental projects may have helped encourage social organization rather than simply resulting from it.

In South America, the stonework of the Inca continues to amaze researchers and visitors. At sites such as Sacsayhuamán near Cusco, enormous stones were shaped and fitted together with remarkable precision.

Some stones weigh dozens of tons, yet they fit so tightly together that a thin object cannot easily be inserted between them. The Inca achieved this without using metal machinery or modern cutting equipment.

Their methods involved careful shaping, testing, and adjustment. Workers likely used stone tools, abrasives, and repeated fitting techniques to create structures that were both beautiful and highly resistant to earthquakes.

This construction style demonstrates a deep understanding of materials. Instead of forcing stones into rigid structures, Inca builders created walls capable of moving slightly during seismic events while remaining stable.

The ancient city of Petra provides another example of construction adapted to difficult conditions. Carved directly into sandstone cliffs, Petra appears almost impossible because its architecture is integrated into the natural landscape.

The Nabataeans transformed a desert environment into a thriving city through engineering and planning. They carved buildings, created water systems, and developed methods for controlling flash floods.

Their greatest achievement was not only the appearance of the city but the infrastructure that allowed thousands of people to live there in a harsh environment.

Ancient temples also demonstrate extraordinary construction abilities. In Cambodia, the Khmer Empire built Angkor Wat, one of the largest religious monuments in the world.

The complex required enormous amounts of sandstone, extensive planning, and sophisticated water management. Engineers designed reservoirs and channels that supported both agriculture and construction efforts.

The precision of the carvings, the scale of the architecture, and the organization required to complete the project reveal a civilization with advanced artistic and engineering capabilities.

Across the Pacific, the moai statues of Rapa Nui (Easter Island) represent another fascinating construction mystery. Hundreds of enormous stone figures were carved, transported, and placed across the island.

For many years, researchers debated how these massive statues were moved. Experimental archaeology has shown that different methods, including the use of ropes, wooden supports, and coordinated movement, could have allowed communities to transport them.

The real achievement was not only moving the statues—it was developing the social organization required to complete such ambitious projects.

Ancient construction mysteries often exist because written records are incomplete. Many builders did not leave behind detailed explanations of their methods. Instead, they left the results of their knowledge in stone.

Archaeologists reconstruct these techniques by studying tool marks, unfinished structures, construction debris, and experimental models. Each discovery provides another piece of the puzzle.

Modern technology has also transformed research. Three-dimensional scanning, satellite imaging, and computer simulations allow scientists to examine ancient structures in ways that were impossible in earlier generations.

These tools reveal hidden chambers, structural patterns, and construction methods while helping researchers preserve fragile sites.

However, the greatest mystery behind ancient construction is not the buildings themselves. It is the human effort behind them.

Every monument represents countless decisions: where to build, which materials to use, how to transport resources, how to train workers, and how to overcome unexpected challenges.

These achievements required cooperation between architects, laborers, craftsmen, engineers, and communities. They were not created by a single person but by generations of shared knowledge.

Ancient construction also teaches an important lesson about innovation. Technology does not always mean advanced machines. Sometimes the most powerful tools are observation, patience, teamwork, and understanding natural materials.

The builders of the past worked with what they had, yet they achieved results that continue to inspire the world.

Their structures remind us that human creativity has no simple timeline. Great ideas can emerge in any era. A civilization does not need modern technology to produce extraordinary achievements—it needs imagination, knowledge, and the determination to turn ideas into reality.

The so-called impossible constructions of history are not evidence that ancient people possessed mysterious abilities. They are evidence of something even more remarkable:

Human beings have always been capable of achieving more than we imagine.

The stones remain. The builders are gone. But their message survives in every monument they created:

The limits of human achievement are often defined not by what is possible, but by what we are willing to attempt.

Chapter 18
Hidden Knowledge Inside Ancient Art

Art is often viewed as a form of beauty, expression, and storytelling. Ancient paintings, carvings, sculptures, and symbols decorate temples, tombs, pottery, and monuments around the world. Yet beneath their visual appearance, many works of ancient art contain layers of knowledge that reveal how civilizations understood science, nature, medicine, astronomy, religion, and the human experience.

For ancient societies, art was not always created simply for decoration. It was a language. It recorded information, preserved traditions, communicated beliefs, and sometimes captured discoveries that were not written elsewhere.

Every symbol, pattern, and image could carry meaning. A carved animal might represent an environmental observation. A geometric design might reflect mathematical understanding. A painting of the sky might preserve generations of astronomical knowledge.

Ancient art serves as a window into worlds that no longer exist.

One of the earliest examples of art containing hidden knowledge can be found in prehistoric cave paintings. For decades, researchers viewed these images primarily as expressions of early creativity. However, further study has revealed that some cave art may contain evidence of careful observation of animals, seasons, and human behavior.

The famous cave paintings of places such as Lascaux in France demonstrate remarkable attention to animal anatomy. The artists understood movement, posture, and physical characteristics with impressive accuracy. Some researchers believe these images may have helped preserve knowledge about hunting strategies, animal behavior, and seasonal patterns.

Although the exact meanings remain debated, one fact is clear: prehistoric artists were not simply decorating cave walls. They were recording their understanding of the world around them.

Ancient Egyptian art provides another extraordinary example of knowledge hidden within imagery. Egyptian paintings and carvings followed specific symbolic systems that communicated information about religion, society, agriculture, and daily life.

Tomb paintings often depicted farming activities, food production, medicine, tools, and astronomy. These images provide historians with valuable information about ancient Egyptian technology and customs.

The artwork inside Egyptian tombs was also connected to their understanding of the universe. The placement of symbols, orientation of structures, and representations of celestial bodies reflected a civilization deeply interested in the relationship between Earth and the cosmos.

In Mesopotamia, art served as a record of power, mythology, and scientific observation. Cylinder seals, sculptures, and relief carvings contained detailed images of rulers, rituals, animals, and symbolic scenes.

Some artwork preserved information about ancient clothing, tools, transportation, and social organization. Because written records from ancient societies are often incomplete, visual evidence provides essential clues about how people lived.

The civilizations of Mesoamerica created some of the most sophisticated artistic records in history. Maya artwork, including carvings, murals, and pottery designs, often contained information about astronomy, calendars, rulers, and ceremonies.

Maya artists represented celestial events and important dates through visual symbols. Their artwork reflected their detailed understanding of planetary movements and cycles of time.

The famous Maya calendar systems were not only mathematical achievements but also cultural expressions. Art helped preserve astronomical knowledge by connecting numbers, symbols, and religious meaning.

Ancient Greek art also contained observations of the human body and natural world. Greek sculptors studied anatomy carefully, creating representations that reflected knowledge of muscles, movement, and proportions.

Their artistic approach influenced later scientific studies of the body. The connection between art and anatomy became especially important during later periods when artists and scientists worked together to better understand human structure.

Beyond realistic representation, ancient art often contained mathematical principles. Patterns found in architecture, pottery, textiles, and mosaics reveal the use of symmetry, repetition, geometry, and proportion.

Ancient artisans understood mathematical relationships even when those ideas were not expressed using modern formulas. They applied these principles through experience, observation, and tradition.

The use of symbols in ancient art also preserved knowledge that could be understood by members of a culture while remaining mysterious to outsiders.

Religious symbols, astronomical signs, and cultural images often carried multiple meanings. A single image could communicate stories, values, historical events, and scientific observations at the same time.

Some ancient artwork has raised intriguing questions because it appears to depict objects or concepts that seem unusual for its time. Researchers approach these discoveries carefully, examining historical context rather than assuming extraordinary explanations.

Often, what appears mysterious becomes understandable when viewed through the knowledge, beliefs, and environment of the people who created it.

Ancient art also provides evidence of lost technologies and skills. The details shown in paintings and carvings reveal tools, construction methods, farming practices, transportation systems, and manufacturing techniques.

For example, artwork showing ships provides clues about ancient navigation. Images of medical procedures reveal approaches to healing. Depictions of machines or tools demonstrate engineering knowledge.

Artists were often observers as much as creators.

One fascinating area of research involves pigments and materials. Ancient artists developed sophisticated methods for creating colors that lasted thousands of years. Scientists studying these materials have learned about ancient mining, chemistry, trade networks, and manufacturing processes.

The creation of a single color could reveal information about where materials came from and how civilizations exchanged resources.

Ancient art also preserved knowledge during times when written records disappeared. When libraries were destroyed or languages were forgotten, images sometimes remained as silent witnesses.

Archaeologists and historians continue to decode these visual messages using technology, comparative studies, and cultural analysis. Advanced imaging techniques can reveal hidden layers beneath paintings, identify original colors, and uncover details invisible to the human eye.

These discoveries show that ancient artwork often contains more information than previous generations realized.

Perhaps the greatest mystery of ancient art is how much remains undiscovered. Every artifact may contain clues waiting to be interpreted. A symbol carved into stone, a pattern painted onto pottery, or a figure hidden within a mural may reveal forgotten aspects of human history.

Art connects imagination with knowledge. It preserves not only what people believed but also what they observed, measured, and understood.

The ancient artists who created these works were historians, scientists, storytellers, and record keepers. They captured pieces of knowledge that survived wars, environmental changes, and the passing of centuries.

Their creations remind us that information is not preserved only in books. It can exist in stone, paint, metal, clay, and symbols.

Ancient art is more than a reflection of the past.

It is a message from the past.

And every generation that studies it discovers something new hidden within the images left behind.

Chapter 19
The World's Most Misunderstood Ruins

Ruins are the remains of civilizations that once existed, thrived, and transformed the world around them. They are often viewed as symbols of mystery—silent stones, abandoned cities, and forgotten monuments waiting to reveal their secrets.

Yet many ruins have been misunderstood for centuries. Some were incorrectly labeled, their purposes misidentified, or their creators overlooked. Others became surrounded by myths that overshadowed the real achievements of the people who built them.

Behind every ruined structure is a story. Not only the story of how it was built, but also the story of the people who lived there, the challenges they faced, and the knowledge they left behind.

The world's most misunderstood ruins reveal a powerful lesson: history is constantly changing as new evidence emerges.

One of the most famous examples is Stonehenge in England. For centuries, people have debated its purpose and meaning. Popular imagination has connected the monument to mysterious rituals, lost civilizations, and unexplained forces.

However, archaeological research has revealed a more complex story. Stonehenge was built through generations of effort by Neolithic communities using careful planning, transportation methods, and astronomical observation.

The massive stones required enormous coordination. Some were transported from distant locations, demonstrating impressive organization and engineering skills. Evidence suggests the site may have served multiple purposes, including ceremonial gatherings, burial practices, and connections to seasonal cycles.

The true mystery of Stonehenge is not who created it or whether it required impossible abilities. The real mystery is how ancient communities organized themselves well enough to create such a monumental achievement.

Another misunderstood site is Machu Picchu in Peru. Often described as a "lost city," it was not actually unknown to local Indigenous communities before its international attention in the twentieth century.

Built by the Inca Empire in the fifteenth century, Machu Picchu is sometimes incorrectly portrayed as an isolated mystery. In reality, it was a carefully designed settlement connected to the larger Inca world.

Its location in the Andes required extraordinary engineering. The builders created agricultural terraces, water channels, and drainage systems that allowed the city to survive in a challenging mountain environment.

Machu Picchu's greatest achievement was not simply its beauty. It was the way the Inca adapted architecture and agriculture to the landscape.

The ruins of Pompeii in Italy have also been misunderstood. The city is often remembered only as a tragedy—a Roman settlement destroyed by the eruption of Mount Vesuvius in 79 CE.

While the disaster preserved remarkable details of Roman life, Pompeii was not simply a frozen disaster site. It was a vibrant city with businesses, homes, political spaces, entertainment venues, and complex infrastructure.

The ruins provide a rare glimpse into everyday life. They reveal how ordinary Romans lived, worked, decorated their homes, managed businesses, and interacted with their community.

Pompeii's importance comes not only from how it ended, but from what it preserved about how people lived.

Another misunderstood ruin is Angkor in Cambodia. Many people know Angkor Wat as a magnificent temple, but the surrounding region was once part of a vast urban civilization.

The Khmer Empire created an enormous network of temples, reservoirs, roads, and water management systems. The city of Angkor was one of the largest urban centers of the medieval world.

For many years, researchers underestimated the scale and complexity of Angkor because much of the city was hidden beneath forests. Modern technology, including aerial mapping, has revealed extensive infrastructure beyond the famous temples.

Angkor was not simply a collection of beautiful buildings. It was an advanced city built around water management, agriculture, and social organization.

The ancient city of Teotihuacan in Mexico presents another fascinating mystery. For centuries, people incorrectly assumed it was built by the Aztecs because they discovered and admired the abandoned city.

In reality, Teotihuacan was created centuries before the Aztec Empire rose to power. Its original builders remain uncertain, but the city's influence was enormous.

Teotihuacan contained massive pyramids, carefully planned avenues, apartment compounds, and complex artistic traditions. At its height, it was one of the largest cities in the ancient world.

The misunderstanding of Teotihuacan demonstrates how later civilizations sometimes inherit the achievements of earlier societies without knowing their full history.

The ruins of Great Zimbabwe in southern Africa were also surrounded by false interpretations. When European explorers encountered the impressive stone structures, some refused to believe that local African civilizations could have created them.

Archaeological evidence later confirmed that Great Zimbabwe was built by African communities between the eleventh and fifteenth centuries. It was a major political and trading center connected to wider regional networks.

Its stone architecture, economic influence, and cultural achievements demonstrate the sophistication of medieval African civilizations.

Another remarkable example is the ancient city of Petra in Jordan. Often called the "rose city," Petra is famous for its buildings carved into sandstone cliffs.

Many visitors see only the dramatic monuments carved into rock, but Petra was much more than a collection of beautiful structures. It was a thriving commercial center created by the Nabataeans.

The city's survival depended on advanced water engineering. The Nabataeans built channels, dams, and reservoirs that allowed them to manage scarce water resources in a desert environment.

Petra was not a mysterious hidden city created by unknown builders. It was the result of human ingenuity and careful planning.

The ruins of Cahokia in North America are another example of forgotten achievement. Located near present-day St. Louis, Missouri, Cahokia was once the largest urban center in pre-Columbian North America.

For many years, the complexity of Indigenous civilizations in North America was underestimated. However, Cahokia's massive earthworks, planned neighborhoods, and population size reveal a highly organized society.

The city demonstrates that monumental construction was not limited to ancient civilizations of Europe, Asia, or the Middle East.

Misunderstood ruins often share a common pattern. They are underestimated because people judge the past through the assumptions of the present. When evidence does not fit familiar ideas, it is sometimes easier to create myths than to reconsider old beliefs.

Archaeology continues correcting these misunderstandings. New discoveries reveal that ancient societies were innovative, organized, and deeply connected to their environments.

Ruins are not mysteries because ancient people lacked knowledge. They are mysteries because time has separated us from the people who created them.

Every stone carries evidence of human decisions: where to build, how to adapt, what to value, and what to preserve.

The world's misunderstood ruins remind us that history is not a finished story. It is an ongoing investigation.

The past is not silent.

It is waiting for us to understand what it has been saying all along.

Chapter 20
Forgotten Scientific Experiments

Science is often remembered through its greatest successes—the discoveries that changed the world, the inventions that transformed society, and the names that became symbols of human achievement. Yet behind every famous breakthrough are countless experiments that were forgotten, abandoned, or misunderstood.

Some experiments failed but revealed valuable information. Others were ahead of their time and only became meaningful generations later. Some disappeared because records were lost, researchers were overlooked, or society was not ready for their discoveries.

The history of science is not only a story of victories. It is also a story of curiosity, uncertainty, and the countless attempts that helped humanity understand the universe.

Many forgotten experiments reveal an important truth: progress does not always come from the first person who finds the correct answer. Sometimes progress comes from those who ask unusual questions and explore possibilities others ignore.

One of the earliest examples of forgotten scientific investigation comes from ancient Greece. Long before modern laboratories existed, Greek thinkers conducted observations and experiments to understand nature.

The philosopher and engineer Archimedes became famous for his work in mathematics and physics, but many ancient experiments and mechanical designs created by early scientists were lost over time. His studies of buoyancy, motion, and engineering principles influenced later science, yet countless other ancient investigations disappeared because they were never preserved.

Ancient scholars often worked without the benefit of modern technology, relying on careful observation and logical reasoning. Their experiments formed the foundation of scientific thinking.

During the medieval period, scientific knowledge continued developing in ways that are often overlooked. Scholars in the Islamic world conducted experiments in optics, medicine, astronomy, and chemistry.

The scientist Ibn al-Haytham, known in the West as Alhazen, performed groundbreaking investigations into vision and light. His experimental approach challenged earlier beliefs about how sight worked.

His work demonstrated the importance of testing ideas through observation rather than accepting assumptions. Although his contributions were significant, many similar experiments from the era were not widely recognized outside scholarly circles for centuries.

The history of chemistry contains many forgotten experiments because early researchers often worked in the uncertain space between philosophy and science.

Alchemy is frequently dismissed as a search for impossible transformations, but many alchemists conducted careful experiments with minerals, metals, and chemical processes. Their work contributed to laboratory techniques, equipment design, and observations that later became part of modern chemistry.

Some experiments failed because researchers misunderstood the nature of matter. However, failure often produced knowledge. Discovering what did not work was an essential part of scientific progress.

In the seventeenth century, scientists began exploring electricity through experiments that seemed mysterious at the time. Researchers investigated static electricity, magnetism, and electrical forces using handmade devices.

Many early experiments were forgotten because they did not immediately lead to practical applications. However, these investigations eventually contributed to the development of modern electrical science.

One fascinating example involves early attempts to create artificial life or understand biological processes. Scientists across different periods experimented with fermentation, microorganisms, and spontaneous generation—the idea that life could arise from nonliving matter.

Some of these experiments produced incorrect conclusions, but they encouraged researchers to develop better methods for studying life. Over time, repeated testing helped replace assumptions with stronger scientific explanations.

The nineteenth and early twentieth centuries produced many overlooked scientific experiments as researchers explored new fields such as psychology, physics, and biology.

Some experiments were forgotten because they challenged popular beliefs. Others disappeared because the researchers who conducted them lacked recognition, funding, or access to influential institutions.

The development of genetics provides a powerful example. Before DNA was understood, many scientists conducted experiments involving inheritance, variation, and biological traits. Some discoveries were ignored until later research provided the missing context.

Gregor Mendel's work on pea plants is one of the most famous examples of a discovery that was initially overlooked. His experiments in inheritance patterns were published in the nineteenth century but received little attention at first. Decades later, scientists rediscovered his research and recognized its importance as a foundation of genetics.

Mendel's story demonstrates that a scientific discovery can exist long before the world understands its significance.

Astronomy also contains forgotten experiments and observations. Ancient and early modern astronomers carefully recorded celestial events, unusual objects, and planetary movements.

Some observations remained unnoticed until later scientists examined historical records and found valuable information hidden within them. Old measurements sometimes provided evidence of long-term changes in stars, planets, and Earth's environment.

The field of medicine contains many forgotten experiments as well. Throughout history, physicians tested treatments, studied diseases, and investigated the human body.

Some experiments were unsuccessful, but they contributed to improved understanding. Others produced important results that were ignored because they challenged accepted medical beliefs.

The development of vaccines, surgical techniques, and disease prevention methods depended on countless observations and experiments conducted over generations.

Modern science continues to rediscover forgotten experiments using new technology. Historical documents, preserved specimens, laboratory notes, and archived data can reveal insights that were impossible to recognize at the time.

Advanced analysis allows researchers to revisit old experiments and ask new questions. A result that seemed insignificant decades ago may become valuable when viewed through modern scientific understanding.

Forgotten experiments also remind us that science is a human process. Researchers make mistakes, change their minds, and sometimes pursue ideas that do not lead where they expected.

The image of science as a straight path toward truth does not reflect reality. Scientific knowledge grows through exploration, debate, correction, and persistence.

Every experiment contributes something, even if the result is not what researchers hoped to find. A failed experiment can reveal a mistake in an idea, a limitation in a method, or a new question worth investigating.

The forgotten experiments of history are valuable because they show the creativity behind discovery. They reveal scientists as explorers willing to enter unknown territory without knowing exactly what they will find.

Some experiments changed the world immediately. Others waited patiently for future generations to recognize their importance.

The greatest discoveries are not always the ones that receive attention first.

Sometimes knowledge is hidden in forgotten notes, abandoned laboratories, and unanswered questions—waiting for someone curious enough to look again.

Chapter 21
The Rise and Fall of Lost Kingdoms

Throughout history, kingdoms have risen from humble beginnings to become centers of power, wealth, and influence. They built magnificent cities, developed advanced technologies, created art and literature, and connected distant regions through trade and diplomacy.

Yet even the greatest kingdoms eventually faced decline.

Some disappeared through conquest. Others faded because of environmental changes, economic struggles, political instability, or shifts in trade. Many left behind only fragments of their existence—ruins, artifacts, inscriptions, and stories passed through generations.

The rise and fall of lost kingdoms reveals a powerful pattern of human history: civilizations are never permanent. Their survival depends on adaptation, leadership, resources, and the ability to respond to change.

The forgotten kingdoms of the past were not failures. They were achievements that existed for centuries before circumstances transformed their worlds.

One of the earliest lost kingdoms emerged in the ancient region of Mesopotamia. The Sumerian city-states created some of humanity's first urban societies, developing writing, organized government, mathematics, and complex religious institutions.

Cities such as Uruk and Ur became centers of innovation and commerce. Their rulers built impressive temples and managed large agricultural systems supported by irrigation.

However, the same systems that created prosperity also created vulnerabilities. Competition between city-states, changing river patterns, political conflicts, and invasions weakened Sumerian power.

Although the Sumerian kingdoms eventually declined, their achievements influenced later civilizations. Their writing system, administrative methods, and technological innovations became foundations for future societies.

Another powerful kingdom that disappeared into history was the Hittite Empire of Anatolia. At its height, the Hittites controlled a significant portion of the ancient Near East and competed with Egypt for regional influence.

They developed advanced military technology, diplomatic systems, and ironworking techniques. Their empire maintained complex relationships with neighboring kingdoms through treaties and political agreements.

Around the end of the Bronze Age, the Hittite Empire collapsed along with several other major civilizations. The causes remain debated, but factors may have included invasions, economic disruptions, internal struggles, and environmental pressures.

For centuries, the Hittites were almost forgotten. Their existence was rediscovered through archaeological discoveries and the translation of ancient inscriptions.

The Kingdom of Kush, located south of Egypt in modern Sudan, is another example of a civilization that was once powerful but later overlooked.

Kush developed wealthy cities, controlled important trade routes, and became known for skilled craftsmanship, architecture, and military strength. At one point, Kushite rulers conquered Egypt and established a dynasty that ruled as pharaohs.

Despite its achievements, Kush was often overshadowed by its northern neighbor. Over time, changes in trade routes, political pressures, and regional competition contributed to its decline.

Today, archaeological discoveries continue revealing the sophistication of Kushite civilization and challenging older assumptions about African history.

The ancient kingdom of Aksum in East Africa also created a remarkable but often forgotten empire. Located in present-day Ethiopia and Eritrea, Aksum became a major trading power connecting Africa, Arabia, India, and the Mediterranean world.

Aksum produced its own coins, developed monumental architecture, and controlled important commercial routes. Its influence extended far beyond its borders.

The kingdom's decline resulted from multiple factors, including changing trade networks, environmental challenges, and competition from rising regional powers.

Although Aksum faded from political dominance, its cultural and religious influence continues today.

In Southeast Asia, the Khmer Empire created one of the most impressive kingdoms of the medieval world. Centered around Angkor, the empire built enormous temples, advanced water systems, and a sophisticated urban landscape.

At its height, Angkor was a thriving center of agriculture, religion, and administration. Its engineers created reservoirs and canals that supported a large population.

However, the empire eventually faced challenges from changing climate conditions, political conflicts, and shifting trade patterns. The decline of Angkor was not a sudden disappearance but a gradual transformation.

The ruins that remain today provide evidence of a civilization that achieved extraordinary accomplishments before adapting to a changing world.

In the Americas, the Maya civilization demonstrates another fascinating example of rise and transformation. The Maya built powerful city-states with advanced knowledge of astronomy, mathematics, architecture, and writing.

Cities such as Tikal, Palenque, and Calakmul became centers of culture and political power.

The decline of many Classic Maya cities has often been misunderstood as a complete disappearance of the Maya people. In reality, Maya communities continued to exist, preserving traditions, languages, and cultural practices.

The collapse was primarily the decline of certain political systems and urban centers rather than the disappearance of an entire civilization.

Another lost kingdom that captured modern imagination is the Nabataean Kingdom. Known for the remarkable city of Petra, the Nabataeans controlled valuable trade routes connecting Arabia, the Mediterranean, and surrounding regions.

Their wealth came from commerce, especially the movement of spices, incense, and luxury goods.

The Nabataeans developed impressive water engineering systems that allowed them to thrive in a desert environment. However, changing trade routes and incorporation into the Roman Empire gradually reduced their independence.

Their legacy survives through Petra and the archaeological evidence of their engineering abilities.

Across history, lost kingdoms often followed similar patterns. They rose because they discovered opportunities—fertile land, valuable resources, strategic locations, or successful trade connections.

They declined when those advantages changed or when societies failed to adapt.

Environmental challenges played a major role in many collapses. Droughts, floods, soil exhaustion, and changing climates could place enormous pressure on civilizations dependent on fragile systems.

Economic changes were equally important. A kingdom built around a particular trade route could suffer when new routes emerged or markets shifted.

Political instability also weakened many kingdoms. Internal conflicts, leadership struggles, and unequal distribution of resources could gradually damage even powerful societies.

Yet the fall of a kingdom did not mean the end of its influence.

Ideas survived. Technologies spread. Languages evolved. Artistic traditions continued. The achievements of lost kingdoms became building blocks for future civilizations.

Modern archaeology continues uncovering these forgotten worlds. Each excavation, inscription, and artifact adds another piece to the human story.

Lost kingdoms remind us that history is not simply a record of winners and losers. It is a record of creativity, adaptation, struggle, and transformation.

The ruins of ancient kingdoms are not just evidence of what disappeared.

They are evidence of what humanity accomplished.

Every fallen kingdom leaves behind a message: greatness can be temporary, but the knowledge created along the way can last for thousands of years.

Civilizations may vanish from maps, but their discoveries continue shaping the world long after their rulers are gone.

Chapter 22
The Ancient Origins of Modern Inventions

Modern technology often appears to be the result of recent breakthroughs—created by laboratories, computers, and advanced engineering. Yet many inventions that define the modern world have roots stretching back thousands of years.

Long before electricity, digital technology, and industrial machines existed, ancient civilizations were developing ideas that would later influence modern inventions. They created early versions of systems involving communication, transportation, medicine, automation, engineering, and mathematics.

Many modern technologies are not entirely new creations. They are the result of a long chain of human experimentation, improvement, and rediscovery.

The story of invention is not a straight path from primitive tools to advanced machines. It is a collection of ideas passed between generations, cultures, and civilizations.

Ancient innovators laid foundations that continue to shape the world today.

One of humanity's most important inventions is writing, which transformed communication forever. The earliest writing systems developed in ancient Mesopotamia, where the Sumerians created cuneiform thousands of years ago.

Writing allowed people to record laws, trade agreements, scientific observations, stories, and historical events. Modern communication systems—from books and newspapers to digital documents—depend on the same basic idea: preserving information beyond human memory.

The invention of writing created the foundation for education, administration, and organized knowledge.

Ancient civilizations also developed early forms of computing and information processing. The famous Antikythera mechanism from ancient Greece demonstrates that people more than two thousand years ago could design complex mechanical systems capable of calculating astronomical cycles.

Although it was not a computer in the modern sense, it represented an early attempt to automate calculations through mechanical processes. The principles behind using machines to process information would eventually become central to modern technology.

Mathematics itself became one of humanity's greatest inventions. Ancient civilizations developed numerical systems, geometry, and calculation methods that continue influencing science and engineering.

The Babylonians created sophisticated mathematical techniques for astronomy and measurement. The Egyptians used geometry for construction and land surveying. Greek mathematicians developed theories that remain essential in modern science.

Without ancient mathematical discoveries, modern architecture, engineering, physics, and computer science would not exist in their current forms.

Transportation technology also has ancient origins. The wheel, developed thousands of years ago, transformed movement, trade, and construction.

Although often considered a simple invention, the wheel required advanced understanding of materials, balance, and mechanical advantage. Its influence extends from ancient carts to modern vehicles, machinery, and industrial systems.

Ancient shipbuilders also created innovations that shaped modern transportation. Civilizations such as the Phoenicians, Greeks, and Chinese developed advanced sailing techniques, navigation methods, and ship designs.

Their understanding of wind patterns, currents, and ocean travel allowed people to connect distant regions. Modern global transportation still depends on the same fundamental idea: using technology to overcome geographic barriers.

The origins of modern medicine can also be traced back to ancient discoveries. Early physicians studied anatomy, developed surgical techniques, and created treatments based on observation.

Ancient Egyptian medical texts documented injuries, diseases, and treatments. Greek physicians emphasized careful observation and diagnosis. Indian medical traditions developed surgical techniques and approaches to health that influenced later medicine.

Although ancient medicine lacked modern scientific tools, many early practitioners established principles that continue to influence healthcare today.

Engineering innovations from the ancient world also shaped modern construction. Roman builders developed roads, bridges, aqueducts, and concrete technologies that demonstrated advanced understanding of infrastructure.

Their methods influenced later approaches to urban planning and public works. Modern cities continue relying on concepts first explored by ancient engineers: transportation networks, water systems, and durable building materials.

Water management is another area where ancient inventions remain relevant. Civilizations around the world created irrigation systems, reservoirs, underground channels, and flood-control structures.

The ancient Persian qanat system transported underground water across dry landscapes. The Maya created reservoirs to store seasonal rainfall. The Romans built aqueducts that supplied cities with fresh water.

Modern water engineering continues developing many of the same principles: conservation, transportation, storage, and efficient distribution.

Ancient inventions also influenced modern automation. Greek and Chinese engineers created mechanical devices that performed actions with limited human involvement.

Hero of Alexandria designed machines powered by air pressure and steam. Chinese inventors created mechanical clocks and automated devices. These early experiments demonstrated humanity's long-standing interest in creating machines that could perform tasks independently.

The idea of automation eventually evolved into modern robotics and artificial intelligence.

The origins of modern architecture can also be found in ancient innovations. Builders developed techniques for creating stronger structures, managing temperature, and adapting buildings to local environments.

Ancient societies used natural ventilation, shading, thermal mass, and strategic design to create comfortable living spaces. Many modern sustainable architecture concepts reflect principles that ancient builders understood centuries ago.

Even modern entertainment technology has ancient roots. The desire to create moving images, mechanical displays, and immersive experiences existed long before cinema and digital media.

Ancient theaters, mechanical performances, and storytelling traditions reveal humanity's long history of combining technology and imagination.

The development of paper, printing, and information sharing also demonstrates how ancient inventions evolved into modern systems.

Chinese inventors created paper and developed printing methods that transformed the spread of knowledge. These technologies eventually influenced global education, communication, and scientific advancement.

Many inventions that appear completely modern are actually the result of ideas that were improved over centuries.

The smartphone, for example, combines ancient concepts of communication, writing, mathematics, engineering, and information storage. Modern medicine combines thousands of years of observation and experimentation. Modern cities continue solving challenges that ancient builders faced: water, transportation, housing, and resources.

Innovation is not created in isolation. Every generation inherits knowledge from those who came before.

Ancient inventors were not simply creating tools for their own time. They were contributing to a larger human experiment that continues today.

Their discoveries remind us that progress is built through connection. Ideas travel across borders, cultures, and centuries. A concept created in one civilization may become the foundation for a technology developed thousands of years later.

The ancient origins of modern inventions reveal a powerful truth:

The future is often built from ideas that began in the past.

Human creativity has always existed. The tools have changed, but the desire to solve problems, explore possibilities, and improve life has remained constant.

The inventions of today are not separate from history. They are the latest chapters in a story that began when humanity first started asking one simple question:

"What can we create?"

Chapter 23
How Small Discoveries Changed Humanity

History often celebrates the biggest achievements—the construction of enormous monuments, the creation of powerful empires, and the inventions that appear to transform the world overnight. Yet some of humanity's most important changes began with discoveries so small that they were almost invisible at first.

A simple observation. A minor improvement. A forgotten experiment. A tiny adjustment in technique.

These moments reshaped the course of civilization.

Human progress has rarely been created only by dramatic breakthroughs. More often, it has developed through countless small discoveries made by curious people who noticed something others overlooked.

A better way to preserve food. A new method for planting crops. A small improvement to a tool. A different understanding of the natural world.

These discoveries accumulated over generations until they transformed how humans lived, worked, communicated, and survived.

The history of humanity is a reminder that small ideas can create enormous consequences.

One of the earliest examples of a small discovery changing the world was the controlled use of fire. The discovery itself may have begun with simple observations: lightning creating flames, burning materials producing heat, and certain conditions allowing fire to continue.

At first, fire was only a source of warmth and protection. Over time, it transformed human life.

Cooking made food easier to digest, expanded available nutrition, and changed human diets. Fire allowed people to survive in colder environments, create stronger tools, develop pottery, work with metals, and eventually build advanced technologies.

A single discovery about controlling a natural force became one of the foundations of civilization.

Another seemingly simple discovery was fermentation. Ancient people noticed that certain foods and liquids changed over time, creating new flavors and properties.

This accidental discovery led to the creation of bread, fermented beverages, preserved foods, and important biological knowledge.

Thousands of years later, the study of fermentation contributed to modern microbiology, biotechnology, and medicine. What began as an observation about changing food became a pathway toward understanding invisible organisms and biological processes.

Small agricultural discoveries also transformed human history. Early farmers did not suddenly invent modern agriculture. Instead, they made gradual improvements through observation.

They noticed which seeds produced stronger plants, which soils supported better growth, and which planting methods increased harvests.

Over generations, these small choices transformed wild plants into domesticated crops that could support growing populations.

The development of agriculture changed everything. It allowed permanent settlements, encouraged specialization, created trade networks, and eventually led to the rise of cities.

A simple decision to save better seeds helped create the foundations of civilization.

Another overlooked discovery was the use of natural materials for tools and clothing. Early humans experimented with stone, bone, wood, fibers, and animal materials.

A sharper stone edge may have seemed like a minor improvement, but it changed hunting, construction, and survival. Better tools allowed humans to accomplish tasks more efficiently and develop new skills.

The history of technology is filled with these small improvements. A slightly stronger material, a more effective design, or a more efficient process can completely change what is possible.

The invention of the needle is another example of a small discovery with enormous consequences. Early needles allowed humans to create fitted clothing rather than simply using animal skins as coverings.

This simple tool helped people adapt to different climates, travel into colder regions, and develop new forms of craftsmanship.

A tiny object changed human survival.

The discovery of the wheel is often described as a major invention, but its impact came from continued improvements. The wheel required experimentation with shape, balance, materials, and movement.

Once refined, it transformed transportation, trade, construction, and engineering.

The first wheels were not advanced machines. They were simple solutions to a basic problem: moving heavy objects more easily.

That small idea eventually influenced nearly every form of transportation in human history.

The discovery of writing also began as a practical solution. Early accounting marks used to record goods and transactions gradually developed into complex writing systems.

At first, writing was not created for literature or philosophy. It was a tool for remembering information.

Yet this small invention allowed knowledge to survive beyond individual lifetimes. It made governments, education, science, and history possible.

One person's decision to record information created a system that changed humanity forever.

Scientific progress also depends on small discoveries. Many major breakthroughs began with observations that seemed insignificant.

A researcher noticing an unusual pattern. A scientist questioning an accepted belief. An experiment producing an unexpected result.

These moments often become important only later, when future generations recognize their meaning.

The discovery of microorganisms provides a powerful example. For most of human history, people did not understand that invisible organisms influenced disease, food, and nature.

The development of microscopes allowed scientists to observe a hidden world. A small improvement in magnification opened an entirely new field of knowledge.

The discovery of electricity followed a similar path. Early observations of static electricity seemed like curiosities rather than practical technology.

Over centuries, experiments gradually revealed how electrical forces worked. These small discoveries eventually contributed to modern power systems, communication devices, computers, and countless technologies.

The history of medicine also shows the importance of small discoveries. A new plant-based treatment, a better surgical method, or a careful observation of symptoms could improve countless lives.

Many medical advances began with individuals noticing something unusual and asking why it happened.

Small discoveries often succeed because they solve specific problems. A person does not always need to change the entire world to create lasting impact. Sometimes improving one task, one process, or one understanding is enough.

The accumulation of small improvements is one of humanity's greatest strengths.

Civilizations grow not only through extraordinary genius but through millions of ordinary acts of curiosity.

A farmer selecting better seeds. A craftsman improving a tool. A healer recording observations. A scientist testing a question.

Each contribution becomes part of a larger chain of progress.

Modern society is built upon countless forgotten discoveries. The technologies we depend on, the systems we use, and the knowledge we possess are the result of generations of small improvements.

The greatest lesson from history is that important changes do not always announce themselves.

Some arrive quietly.

A small discovery may appear ordinary in its own time, yet centuries later it may become recognized as a turning point in human development.

The future is often shaped by people who notice what others ignore.

Humanity has always advanced because someone looked at the world and asked:

"What happens if we try something different?"

Sometimes the smallest discoveries create the biggest changes.

Chapter 24
The Lessons Civilization Keeps Forgetting

History is often described as a record of what happened. Dates are written, battles are documented, leaders are remembered, and civilizations are studied. Yet beneath the events and achievements lies another story: the lessons humanity repeatedly learns, forgets, and eventually rediscovers.

Civilizations rise because people solve problems. They decline when they ignore warnings, fail to adapt, or repeat mistakes from the past.

The purpose of studying history is not only to understand what came before. It is to recognize patterns that continue appearing across generations.

Ancient societies faced challenges that feel surprisingly familiar today: environmental change, resource shortages, inequality, political conflict, technological disruption, and the struggle to balance growth with sustainability.

The names of civilizations change. The locations change. The technology changes.

But many of the fundamental challenges remain the same.

One of the greatest lessons repeatedly forgotten throughout history is the importance of respecting the environment.

Many successful civilizations depended on careful relationships with their surroundings. They managed water, protected soil, and adapted farming methods to local conditions.

However, some societies expanded faster than their environments could support. Forests were cleared, soils were exhausted, and resources were consumed faster than they could recover.

The decline of some ancient societies demonstrates that environmental damage can weaken even powerful civilizations. Wealth, technology, and military strength cannot completely protect a society that destroys the systems supporting its survival.

The lesson is simple but often ignored:

A civilization cannot separate itself from nature.

The health of the environment and the success of human communities are connected.

Another lesson history repeatedly reveals is that resources are never unlimited.

Ancient cities flourished because they controlled valuable resources—water, fertile land, minerals, trade routes, and food supplies. These advantages allowed populations to grow and economies to expand.

Yet growth often created new pressures. More people required more resources. More construction required more materials. More production required greater consumption.

When societies failed to manage resources wisely, instability followed.

History shows that prosperity requires responsibility. A society must not only create wealth but also protect the foundations that make wealth possible.

Civilizations also repeatedly forget the importance of knowledge preservation.

Throughout history, libraries, schools, archives, and centers of learning have protected humanity's discoveries. When these institutions were destroyed, countless ideas disappeared.

Ancient texts were lost. Scientific observations vanished. Languages became difficult to understand. Technologies had to be rediscovered.

The loss of knowledge demonstrates that information is one of humanity's most valuable resources.

A civilization is not only defined by what it builds.

It is defined by what it remembers.

Another recurring lesson involves the danger of believing that success will last forever.

Many powerful kingdoms and empires reached moments of extraordinary achievement. They controlled territories, accumulated wealth, and developed advanced systems.

However, success often created confidence that became resistance to change.

Societies that believed they were too powerful to decline sometimes ignored warning signs. They underestimated competitors, failed to adapt, or continued outdated practices long after conditions changed.

History repeatedly demonstrates that adaptability is more valuable than temporary dominance.

The ability to change is one of the strongest indicators of survival.

Civilizations also continue struggling with inequality. Throughout history, societies have experienced divisions between powerful groups and those with fewer opportunities.

When large portions of a population feel excluded, ignored, or exploited, social tensions often increase.

Many civilizations faced internal weaknesses not because they lacked resources, but because those resources were distributed unevenly.

The lesson is not that all societies must be identical. It is that long-term stability requires a sense of shared investment in the future.

A civilization is strongest when people believe they have a meaningful place within it.

Another forgotten lesson is the importance of questioning assumptions.

Many scientific and cultural advances occurred because individuals challenged ideas that were considered unquestionable.

Throughout history, societies have sometimes resisted new discoveries because they conflicted with established beliefs.

Yet progress depends on curiosity and the willingness to examine old ideas.

The civilizations that advanced were often those that encouraged learning, debate, and exploration.

A society that stops asking questions eventually stops growing.

Technology provides another important lesson from history.

New inventions have always changed human life. Ancient societies experienced transformations through writing, metalworking, agriculture, navigation, and engineering.

Modern societies experience similar changes through digital technology, automation, and artificial intelligence.

The lesson from history is not that technology is dangerous or beneficial by itself.

Technology reflects the choices of the people who create and use it.

Every invention requires wisdom, responsibility, and consideration of long-term consequences.

Civilizations also repeatedly forget the importance of cooperation.

Many great achievements were possible because people worked together. Monumental buildings, scientific discoveries, trade networks, and cultural accomplishments required collaboration.

However, conflict has often prevented societies from reaching their full potential.

History shows that competition can encourage progress, but cooperation creates lasting achievements.

Humanity's greatest accomplishments have rarely been created by individuals working alone.

Another lesson is that civilizations are built slowly but can change quickly.

A city may take centuries to develop. A culture may take generations to form. Knowledge may require thousands of years of accumulation.

Yet a few decades of poor decisions can weaken what took centuries to build.

This imbalance reminds us that preservation requires constant effort.

The past also teaches humility.

Every civilization, no matter how advanced, believed it understood the world of its time. Yet future generations often discovered knowledge that earlier societies lacked.

Modern humanity is not separate from this pattern.

The technologies and systems of today may also be viewed differently by future generations.

Recognizing this encourages curiosity rather than arrogance.

Perhaps the greatest lesson civilization keeps forgetting is that progress is not guaranteed.

Advancement requires attention, responsibility, and continuous learning.

The achievements of the past were not inevitable. They were created by people who solved problems, made discoveries, and worked toward something greater than themselves.

The challenges of the future will require the same qualities.

History is not a warning that humanity always repeats mistakes.

It is a reminder that humanity has the ability to learn.

Every generation receives knowledge from those before it. The question is whether that knowledge will be ignored or transformed into wisdom.

Civilizations do not survive because they are the strongest.

They survive because they are willing to remember.

Chapter 25
The Blueprint for Tomorrow

Every civilization begins with an idea.

Before a city rises, there is a vision. Before a discovery changes the world, there is a question. Before a future is created, someone must first imagine what does not yet exist.

Human history is a collection of blueprints—plans created by countless generations attempting to build something better. Some blueprints became reality. Others were abandoned, forgotten, or rediscovered centuries later.

The future is not created from nothing. It is built from the knowledge, mistakes, discoveries, and dreams of those who came before us.

The civilizations of the past left behind more than ruins and artifacts. They left instructions. They showed humanity what happens when curiosity is encouraged, resources are protected, knowledge is preserved, and people work together toward shared goals.

The blueprint for tomorrow is not a single invention or idea.

It is the combined wisdom of thousands of years of human experience.

One of the greatest lessons from history is that innovation begins with imagination. Every invention, structure, and discovery started as an idea that existed only in someone's mind.

The ancient builders who created massive monuments imagined possibilities beyond their immediate circumstances. Scientists who studied the natural world imagined answers before evidence confirmed them. Explorers crossed unknown landscapes because they believed something existed beyond the horizon.

The ability to imagine a different future has always been one of humanity's greatest strengths.

However, imagination alone is not enough. The civilizations that succeeded were those that combined vision with knowledge and action.

The ancient engineers who created irrigation systems did not simply dream of controlling water. They studied rivers, tested methods, and developed practical solutions.

The inventors who transformed transportation did not only imagine movement. They experimented with materials, designs, and mechanics.

The future is built by turning ideas into reality.

Another essential part of tomorrow's blueprint is the preservation of knowledge.

History has shown that civilizations can lose valuable discoveries when information is destroyed, ignored, or forgotten. Ancient technologies disappeared because the knowledge behind them was not always protected.

Today, humanity has more information available than ever before. Yet the challenge remains the same: knowledge must be organized, preserved, and shared.

A civilization's greatest resource is not only what it owns.

It is what it understands.

The future will also depend on how humanity approaches the relationship between technology and nature.

Ancient civilizations demonstrated both the possibilities and dangers of human innovation. They created remarkable systems for agriculture, architecture, and water management, but some also experienced decline when resources were exhausted.

Modern society faces similar choices.

Technology can help solve environmental challenges, improve lives, and expand opportunities. However, progress must be guided by responsibility.

The future cannot be built by ignoring the planet that supports it.

The blueprint for tomorrow must include sustainability—not as a limitation, but as a foundation.

Future civilizations will also be shaped by how they use information. Knowledge has always been powerful, but modern technology has accelerated the speed at which ideas spread.

Communication systems allow discoveries to reach people across the world instantly. Collaboration between different cultures and disciplines has the potential to create solutions that no single group could achieve alone.

The civilizations of the future will likely be defined by connection.

The ability to share knowledge, cooperate across boundaries, and learn from different perspectives will become increasingly important.

Another part of the future blueprint involves understanding human creativity.

Throughout history, people have adapted during difficult circumstances. They survived changing climates, created new tools, rebuilt after disasters, and found solutions to seemingly impossible problems.

Human creativity is not limited to one era.

The same curiosity that inspired ancient inventors continues to inspire modern scientists, artists, engineers, and explorers.

The future will belong to those who continue asking questions.

What can be improved?

What remains unknown?

What problems can be solved?

The greatest advancements often begin with curiosity.

The past also teaches that no civilization is perfect. Every society has achievements worth celebrating and mistakes worth studying.

The future should not be built by copying the past exactly. Instead, it should be built by understanding what worked, recognizing what failed, and creating something better.

History is not a blueprint that provides every answer.

It is a foundation upon which new ideas can be constructed.

Tomorrow's greatest achievements may come from combining ancient wisdom with modern innovation.

Traditional agricultural knowledge may influence future food systems. Ancient engineering principles may inspire sustainable designs. Historical approaches to cooperation and resource management may guide future societies.

The past and future are not separate worlds.

They are connected chapters in the same human story.

The blueprint for tomorrow will also require courage. Every major advancement in history involved uncertainty.

The first explorers did not know exactly what they would discover. The first inventors did not know whether their creations would succeed. The first scientists investigating unknown questions often worked without guarantees.

Progress requires the willingness to explore possibilities beyond what is already known.

The future belongs to those who are willing to build before they have all the answers.

Perhaps the most important lesson from history is that humanity's greatest achievements were never created by perfect conditions.

They were created by people who chose to act despite challenges.

Civilizations have always been works in progress. Each generation adds another layer to the structure of human knowledge.

The discoveries of ancient farmers, engineers, artists, scientists, and explorers became the foundation for the world we know today.

The decisions made now will become the foundation for future generations.

The blueprint for tomorrow is still being written.

Every discovery, every question, every act of creativity adds another line to that design.

The future is not something humanity simply enters.

It is something humanity creates.

The story of civilization is unfinished. The next chapter belongs to those who continue searching, learning, and imagining what comes next.

Because the greatest discoveries of tomorrow may already exist as ideas waiting for someone brave enough to build them.

www.ingramcontent.com/pod-product-compliance
Lightning Source LLC
LaVergne TN
LVHW030727061126
851857LV00019B/2028